The Dynamic Architecture
of a Developing Organism

An Interdisciplinary Approach
to the Development of Organisms

by

L. V. BELOUSSOV

KLUWER ACADEMIC PUBLISHERS
DORDRECHT / BOSTON / LONDON

A C.I.P. Catalogue record for this book is available from the Library of Congress.

ISBN 0-7923-5044-8

Published by Kluwer Academic Publishers,
P.O. Box 17, 3300 AA Dordrecht, The Netherlands.

Sold and distributed in North, Central and South America
by Kluwer Academic Publishers,
101 Philip Drive, Norwell, MA 02061, U.S.A.

In all other countries, sold and distributed
by Kluwer Academic Publishers,
P.O. Box 322, 3300 AH Dordrecht, The Netherlands.

Printed on acid-free paper

Printed in the Netherlands

...A palm of primacy will be gained by that most happy one who succeeds in reducing the formative powers of the animal organisms to the general forces or vital laws of a World's Entity.

Carl Ernst von Baer (1828)

We can now already speak about a body as about a mobile border between the future and the past, about a shifted point which behaves as if being permanently pushed by our past towards our future.

Henry Bergson (1896)

According to our viewpoint... out of an adequate description of any momentary state of a living system its transition to the next should inevitably follow.

Alexander Gurwitsch
(written in 1948 - 1950, published in 1991)

A point of primacy will be regarded by the
most happy one who succeeds in reducing the
formative powers of the animal organisms to the
general forces or vital laws of ...

Carl Friedrich von Baer (1834)

We can now already a body as about
a mobile barrier between the future and the past
about a shifted point which behaves as if being
permanently pushed by our past toward our
future

Henry Bergson (1896)

According to our viewpoint ... out of an adequate
description of any mathematical theory ...
applied to matters to the new should therefore
follow

Alexander Carson...

TABLE OF CONTENTS

INTRODUCTION

For anybody capable of an emotional response to it, any view of a developing organism should give birth to a feeling of amazement and even admiration, whether this development is seen directly, or in the form of a time lapse film, or even if mentally reconstructed from a series of static images. We ask ourselves how such seemingly primitive eggs or pieces of tissue, without any obvious intervention from outside, so regularly transform themselves into precisely constructed adult organisms. If we try to formulate what amazes us most of all about development, the answer will probably be that it is the internal capacity of developing organisms themselves to create new structures.

How, then, can we satisfy our amazement in ways that are more or less reasonable, as well as scientifically valuable?

This depends, first of all, on what position we choose to regard embryonic development as occupying among other structure creating processes, even including human activities. On the one hand, one might regard the development of organisms as a highly specialized class of processes, unique to themselves and alien to the general laws of nature, or at least not derivable from them and more akin to the deliberate acts of our own human behaviour. In that case our task would become reduced to a search for some specific 'instructions' for each next member of such a class. Whether in an overt or hidden form, some such ideology seems to dominate in present day developmental biology. On the other hand, meanwhile, a very different approach is possible, one which was suggested by the first of the quotations, that of von Baer who was one of the deepest thinkers in embryology, more than a century and a half ago. This alternative approach regards the formative powers of organisms as manifestations of a much wider set of structure creating activities governed by some universal laws to which both the living and the non-living parts of nature would be subject. As we hope to show in this book, our period of scientific history is probably the first in which such ideas can be converted into a set of definite research tasks, rather than remaining as mere romantic notions or philosophical maxims. And we shall begin this conversion by comparing and evaluating the broadest approaches used for comprehending and interpreting structures and the activities that generate them.

Within the past few centuries the human mind has elaborated two extreme alternative ways for satisfying, as far as possible, the amazement that can result from the perception of new structures, both natural and man made ones. The first of these two alternatives can be called the aesthetic, or holistic, approach. This is the one which we amateurs instinctively use when perceiving either some work of creative art (whether it is a poem, a piece of music, or an architectural edifice) or remarkable natural events. As a rule we perceive such objects as fundamentally indivisible entities, which only specialists would have any interest (also quite carefully and, of course, only mentally) in dissecting into components. We know also that the main and most attractive property of any real artistic creation (as well as a natural object) is a non-predictability of each spatial and/or temporal detail, its non-derivability from its neighbouring and/or a previous parts, with this non-predictability even being increased with its further mental dissection. Another way of defining such non-derivability is to claim that objects of aesthetic value have a property of *freedom*. Certainly, such a freedom is never absolute: all artistic creations (not to say about natural events) are characterised by certain prohibitions. But these can never be codified in an absolutely strict manner (otherwise they would no longer belong to art) and they cannot be recognized by decomposing an object into pieces that are too small. Therefore, such a dissection should be qualified as useless from the heuristic point of view, and perhaps even a sort of crime. Much more can be gained, instead, by enlarging our scope, both in space and in time, for including into it what may be called the *context* of an observed object, that is, the set of other more or less similar objects, their historical or natural background, etc..

It is obvious, at the same time, that the role of the observer's (a listener's, a reader's) own individuality should be, within such an approach, quite important, active, and unique. No artistic creation has any value at all in the absence of those who perceive it adequately, and this value is changed from one person to another. You, my reader, have your own Beethoven symphony or Rembrandt picture, and I have another. A set of associations born in different persons which observe the same landscape may vary immensely. Within the act of such a perception, the artistic or natural creation and the perceiving individual are really fused together into a unique and indivisible whole.

Quite another approach is used while dealing with the structures of any origin which we do not endow by any aesthetical, or holistic, qualities. Whether such structures fulfil or not some clearly outlined functions the usual, accepted way of satisfying our curiosity is now associated as a rule with a decomposition or a dissection of the structures into ever smaller elements, both spatially and temporally. By doing this we assume that the decomposition brings us towards recognition of such strict, unambiguous, and hence inevitable, relationships between the neighbouring elements or the successive events, which will permit us to 'derive' the origination of each spatial and/or temporal element from the preceding and/or neighbouring one. Correspondingly, we now consider the object of our interest as belonging to the 'inevitability world', in the sense that any traces of 'freedom' in its behaviour come to be regarded merely as reflections of deficiencies and gaps in our knowledge.

A principle justification for such an approach, which we can call analytical, is belief in the existence of a one-to-one cause–effect correspondence between events which are temporarily and/or spatially adjacent. This is the central to a very influential epistemological principle, that of a uniform determinism, which will be discussed later on in more detail. We would now merely like to point out that by following such an approach the role of the observer becomes reduced to that of a skilled 'decomposer'. To the extent that we believe the result of a successful decomposition may be only one, any manifestations of an observer's individuality are, or course, rendered undesirable.

But which of these alternative approaches, the aesthetic or the analytical, is most appropriate to the task of understanding the formation of structures by the developing organism? Over the last one or two centuries the answer has seemed quite obvious to most naturalists. "Nature is a workshop, rather than a temple, and a man is a worker in it" – as the Russian plant physiologist Clement Timiryazev quite straightforwardly expressed it. Thus, any aesthetical criteria have been left very much behind by science, which has become almost completely identified with the analytical approach.

Meanwhile, today we know, that the question looks far from simple. It is one of the main aims of this book to find a proper balance between the two trends. As a starting point we would like to remind ourselves that within centuries human beings employed, with a considerable success, a way that can be characterised as intermediate between the two extremes above described.

This was the approach used by a person discovering a new land. On the one hand, in no way could he be satisfied simply by 'apprehending it as a whole' – at no one moment was he able to observe more than a small part of it. But, on the other hand, he had neither the desire nor the physical possibility of 'decompose' it. What he actually did was to describe, step by step, some of the more or less remarkable details of the landscape: sea shore; river estuaries; the most prominent hills; etc.. Soon he realised that he must do two different, although related, things: first, select a scale (or a set of scales) for his description, and, second, establish reference point(s). This is a crucial moment, which, on the one hand, calls for some arbitrary individual decisions, but, on the other hand, requires good knowledge of the properties of the land explored. Consequently freedom of choice becomes intimately entangled with inevitability, as they do in many cases of further investigation. To select a scale means deliberately to ignore some structural details in favour of others, these becoming recorded as precisely as possible. But that need not prevent us from passing later to some other scale of observation, in which some other sets of elements will be targeted. In such ways we apprehend what is called a *hierarchical structure* of the object described. This raises entirely new questions about the inter-relations between elements at different scales: a priori, these cannot be reduced to one-to-one inevitable connections. Consider, for example, a river valley. The height of its banks depends, at one level, upon the molecular (crystal) properties of the rocks (in other words, depending on the events of a microscopic scale). At other levels, however, the valley's orientation relative to the earth's axes, amongst many other macroscopic properties, also affects the heights of the banks. In both cases, however, the dependence is far from being as strict as is demanded by uniform determinism.

A next crucial moment in the investigations comes when a discoverer realizes that that land not only has spatial, but also temporal structure, in other words a *history*. This is a transition from a uni-temporal geography to a multi-temporal geology. Comparable to the linear scales mentioned before, we will now find a hierarchy of temporal scales, and will discover that both are, as a rule, parallel to one another. Thus, some small components of a river valley can change their arrangement within a day or so; larger components exhibit seasonal changes; and even greater ones require climatic changes occurring over several centuries. Finally we come to the geological time scale. In such a way we have to consider our land as a set of loosely interconnected *dynamic*, rather than static, structures; it is the totality of such structures that we may define as the *dynamic architecture* of the land. With such knowledge in hand we can make some predictions concerning the 'creative activity' of this land: its availability for establishing factories (or temples, contrary to Timiryazev's claim), or for developing agriculture, or navigation, or for predicting earthquakes. Note, meanwhile, that the things we predict are no more than possibilities, or *potencies,* rather than something inevitable. Again, a component of freedom cannot be completely avoided. A potency itself is a notion with a very deep meaning which we will begin to discuss in Chapter 1.

Lastly, we reach in our analysis of the dynamic architecture a rather dangerous point. While studying carefully and with the use of special equipment, for example, a water dynamics within a river valley, we come sooner or later to the conclusion that the river itself, the atmosphere above it, and the soil around it are far from being completely isolated entities; instead, they create a certain indivisible whole involved in the water turnover together with a number of other substances. Does this mean that our structural analysis was in vain? Such a conclusion would also be premature. A proper answer should be that the units which we have distinguished from each other are not, in fact, completely different and isolated entities. Instead, they should be qualified, in physical terms, as the *singularities of a common dynamic field.* Or, using the expression of David Bohm, an outstanding thinker of the recent past, "all is an unbroken and undivided whole movement... each thing is abstracted only as a relative invariant side or aspect of this movement" (Bohm, 1983).

Let us now ask ourselves whether there really is much in common between a land surveyor's approach and that which would be largely adequate for the purposes of studying a developing organism. There is more, than it may seem at first glance. About half a century ago the British embryologist Conrad Waddington (1940) used the analogy of a mountain landscape (which he called the 'epigenetic landscape') for the purpose of representing his views. What he sought to represent by this analogy was exactly what we have called the dynamic architecture of development. Although this type of analogy has unfortunately failed to attract many proponents, we will try to extend this approach further.

So far as we will be interested in the dynamic, which is to say the temporally extended, structures their time component will be of special interest to us. Within each time scale we will consider the history of an element (its more or less prolonged past) as a bearer of its active memory, 'pushing – in Bergson's terms, see the second quotation above – an element towards its future'. In such a view, each moment of developmental time should be extended into the past (transformed into a finite 'metamoment', see Anisov, 1992) in order for its future to become at least loosely predictable.

The plan of this book is as follows. We will start in Chapter 1 by reviewing briefly the main approaches used by the human mind for comprehending the formation of both non-living and living structures in nature. We will follow the tortuous path of the evolution of such approaches in physics and biology which, as I hope, are at last converging towards one another after centuries of a mutual neglect and ignorance. Some ideas and notions recently developed within the social sciences will also be employed.

After such an excursion in Chapter 2 we will return to biology in order to outline and describe one after another the dynamic structures of the different spatio-temporal levels. Next we will explore their interactions, paying special attention to the possibilities of establishing feedbacks between these different levels. This will permit us to make some generalizations formulated in the terms of a 'morphomechanics'. This will bring us closer to our ultimate task: to examine how the dynamic structures of the different scales become integrated into a coherent multilevel hierarchy, which we will define as the dynamic architecture or, in a more operational manner, as the developmental successions. Accordingly, the main aim of the Chapter 3 will be to explore a way for reconstructing the developmental successions in the different groups of organisms.

The construction of the book permits a reader to select, according to his or her interests, some parts or chapters at the expense of others. Those who do not belong to the developmental biologists proper but are interested in the interdisciplinary connections and in the general scientific trends may restrict themselves by reading the first part of the book and the concluding remarks, missing all the rest. The second and the third chapters are, in contrast, more special. Yet these two parts, however, I have tried to construct in such a way as to make their qualitative conclusions as far as possible independent from mathematical formalism (which is minimized to the greatest possible extent).

The process of writing this book has itself been for me kind of a 'temporally extended moment' of talking and discussing questions in my imagination with a number of persons, some of whom died as long as several decades ago, as well as with others who fortunately remain alive and active. Inspite of not sharing all their points of view, I am nevertheless indebted to them in everything. The person to be mentioned first of all should be my teacher Alexander Gurwitsch (1874-1954) who in my view was one of the greatest biological thinkers of the twentieth century, unfortunately too little known outside Russia. Amongst others, who have left us over the years I would like to remember Professors Alexander Liubishev, Pavel Svetlov, Vladimir Beklemishev, Anna Gurwitsch and the prematurely deceased brilliant researchers, Boris Belintzev and Vladimir Mescheryakov. And, finally, I would like to name the persons of a different age, whom I am happy to regard as my scientific friends, and who have enriched me so much by their ideas and views (but in no way are responsible for any possible shortcomings of this book). A very incomplete list of them includes Wolfgang Alt, Juergen Bereiter-Hahn, Vladimir Cherdantzev, Dmitry Chernavsky, Richard Cummings, Joe Frankel, Brian Goodwin, Albert Harris, Michael Lipkind, Jay Mittenthal, Fritz Popp, Guiseppe Sermonti and Alexander Stein. Amongst those, my special and deep thanks go to Albert Harris, who took a burden to look through a great part of the book, both from a linguistic and scientific point of view, and to Fritz Popp, an enthusiastical founder of the International Institute of Biophysics to which I am happy to belong. I would like also to express my thanks to my colleagues at the Faculty of Biology and, in particular, the Laboratory of Developmental Biophysics, Moscow State University, for maintaining an atmosphere of a mutual support and benevolence during all the time we have spent together.

CHAPTER 1
STRUCTURE FORMATION IN PHYSICS AND BIOLOGY: GENERAL OUTLINE AND APPROACHES

1.1. From the spontaneous creativity of Nature to the inert world of uniform determinism (an historical outlook).

As we are taught by the mythologies of quite diverse peoples, the world of primitive man was full of a spontaneous creativity, sometimes benevolent, but at other times alien and dangerous. Primitive peoples considered their own activities as no more than extensions of those exhibited by nature around them, both animate and inanimate. The elements of an Order, that is of positive creation of some definite and recognizable space-time structures, were tightly intermingled, in their conceptions, with inevitable bursts of Chaos, that is of a fatal and unavoidable destruction, with a loss of the structure's recognizability. Such a direct and intuitive appreciation of the surrounding world as creative and chaotic at the same time passed from mythology to ancient philosophy, with one of its highlights exemplified by Aristotle's conceptual system (see, e.g., Aristotle, 1975).

One of its central ideas was that natural bodies are able to undergo much more profound transformations than mere changes in their positions or sizes. These transformations were, from a philosophical point of view, of a qualitative rather than quantitative nature, being owed, as a rule, to an inherent spontaneous activity of the bodies themselves, rather than to external influences upon them. Without denying these influences completely, Aristotle nevertheless restricted their role by providing active bodies with either a raw material ('causae materialis') or with something more or less resembling the modern notion of an energy (a heat, required for forging a sword or for incubating a hen's egg) – those were 'causae efficiences'. Of a much greater importance, meanwhile, were, in Aristotle's view, two other causes, very much different from the previous ones and belonging to the natural bodies themselves (or, in the case of artifacts, to a man-creator). Such was the 'causa formalis', or a form proper, to which an evolving body have striven, and the 'causa finalis', a purpose of a body's existence.

This famous Aristotelian tetrahedron of causes, so mercilessly withdrawn from the Modern Age science by Francis Bacon and his followers, today attracts a renewed interest and raises wide associations. In a modern terminology it is a multilevel and time extended construction: while the first two causes well may act locally and instantaneously, the two latter are inevitably related to the whole body, rather than to its separate parts. What is even more remarkable, they should act from the future towards the past. Therefore, the first scientific attempt at apprehending a structuring activity of Nature linked itself with what can be defined as a *temporal depth*, although the latter have been projected from the present time moment towards the future, rather than coming towards the present from the past (which is more common for the modern science). In any case, there is no doubt that Aristotle definitely accepted the creation of actually new structures from a less structured, or a completely structureless, state. Without denying completely the influences of the preceded structures on the succeeded ones, the philosopher considered these influences as no more than a slight "touching": this was an excellent allegory, which we shall recall later in this book. In general, he did not consider a succession of embryonic shapes as something born by 'necessity'. In the other words, he allowed a substantial amount of a freedom in the structure formation processes.

Such a beautiful philosophy had but one fatal deficiency: it could not provide any predictions, in particular ones of a quantitative nature. This was the main cause of its complete rejection since the beginning of the Modern Age. For the sake of a quantitative predictiveness, what had seemed a unified World has been now radically split into separate realms of freedom and of necessity – with the human beings as sole inhabitants of the former one. Not just non-living matter, but likewise all animate creatures besides man have been deprived of any kinds of spontaneous activity, except as induced from outside. Nature came to be viewed as fundamentally passive and inert, with each movement requiring some external impulse, or a force: Descartes' reflexes, Newton's laws of mechanics, and Leibnitz' philosophy of a constant world with nothing new ever happening were among the first milestones of this new and, by all accounts, triumphant march of science.

As a result, this 'world of necessity', the only one qualified as deserving a truly scientific approach, appeared to be divided into short space-time fragments: at the beginning of each of these was a 'cause', or, alternatively, some 'initial (or boundary) conditions', while at the end was the effect, or result, of the process. Such cause-effect relations have been considered as strict and uniform. One cause could produce no more than one effect, although the converse was not demanded, in that certain effects could require more than one cause: both causes and effects had to be strictly localised; and they ought to have a common measure in the sense of being quantitatively comparable with each other – that is, too 'small' causes could not produce too 'large' effects, and vice versa. Such were some of the postulates of an ideology of uniform determinism, overtly explicated by François Laplace about century after the formulation of Newton's laws.

According to this ideology, our world could, in principle, be completely explained backwards into the past, as well as completely predicted in the forward direction into the future; it is only the incompleteness of our present day knowledge which leaves the place for our partial ignorance. 'If we imagine a mind that could know at one given time all the relations between the World's objects', wrote Laplace in 1776, 'it will be able to establish the corresponding positions, motions and the mutual influences of all these objects at any time in the past or the future''.

The world of classical determinism was completely reversible: the future and the past reflected one another in a one to one manner, so that nothing really new and irreversible could ever happen. Historically, this idea was the basis for the great laws of conservation (of momentum, energy and mass). Meanwhile, it was just the postulate of a time reversibility that was undermined by a subsequent development of physics; in a rudimentary form this has been already done by classical thermodynamics. For a long time period, however, the thermodynamical irreversibility of time strangely coexisted with a continuing acceptance of a time reversibility in mechanics.

Another highly influential methodological idea, also directly emerging from the principles of uniform determinism, was a so called reductionism or, more exactly, a micro-reductionism. This is an empirical imperative, directed towards splitting a studied object into ever smaller spatio-temporal constituents in the hope of establishing, at some time moment, a mostly strict and uniform cause – effect connection between its components. Paradoxically, this ideology, being of a physico-mathematical origin, never played in these sciences a role to any extent comparable with its really demoniacal intervention in biology. Within a century at least, for most investigators progress in the biological knowledge has been completely identified with a continuing subdivision of an object into diminished components.

On the other hand, a much more important, and heuristically most useful, methodological idea of a Modern Age science appeared to be much less explicated and appreciated. What we mean is a pursuit that considers the laws of a Nature as mathematical *invariants,* preserving their structure within as broad as possible changes of the variables. It was this ideology which created a basis for a symmetry theory, indispensable for interpreting the formation of all the kinds of structures. We now begin to trace this line.

1.2. Non-biological structurization in terms of a symmetry theory.

A theory of symmetry looks, at the first glance, or at least in its most elementary aspects, like a set of very simple and almost trivial constructions. However, it is actually based upon very deep ideas and is elaborated in modern science up to highly refined mathematical constructions. According to one of the most competent definitions (Weyl, 1952) it is a science about invariant transformations, that is, about transformations which either do not disturb the self-identity of an entire body or, at least, do not alter some of its important characters. As it has already been mentioned, the pursuit of invariance is an ultimate goal of modern science; a symmetry theory may thus be considered as a kind of schematised prototype of it.

Both the transformations and the characters which should be retained invariant may be quite different. In elementary symmetry theory the transformations are reduced to the movements of rotation, reflection, and translation (the latter are progressive linear shifts). We take the liberty of adding to this list two typically biological procedures associated with the shifts of embryonic material: (1) *transplantations,* which formally are the 'jumps' of some definitely outlined areas of an embryo from one of its regions to another; and (2) *explantations,* which are similar shifts to outside the embryo.

Before considering what kinds of body characters can remain invariant during these transformations we must become familiar with the different kinds of spaces into which these transformations can take place. The first of them is the usual *physical* (Euclidean) space, which may be one-, two- or three-dimensional. In addition, we can speak about a temporal dimension, either taken isolated from a physical one (in these cases we will be dealing with a purely temporal symmetry), or combined with a 3-dimensional physical space into a 4-dimensional space-temporal continuum. And, finally, symmetrical transformations may take place in a so called *phase space,* which is a space of certain properties (variables), plotted along the co-ordinate axes instead of the linear or temporal units. A phase space may be of any dimensionality, depending upon the number of the variables to be accounted for. Mixed phase-physical spaces are also possible.

As we have already said, the strongest demand on the invariance of the symmetry transformations ('strong' symmetry) is in keeping the identity of a body explored. In the other words, after a given set of transformations a body should be indistinguishable from the same body prior to the transformations.

In their turn, two kinds of identity can be formulated. In the simplest case we assume that all of the local areas of a given body, up to negligibly small ones ('points') are indistinguishable from each other. Such a body can be defined as a mono-coloured one. In this case the only criteria for identifying a body is its shape proper, i.e., its geometry. The corresponding kind of symmetry is called *geometrical*. It will be fulfilled if, after certain transformations, a body coincides with itself by superposition.

Another thing would be if at least some small areas (points) of a body are somehow distinguished from their surroundings. We may describe these points as coloured. Now the symmetry conditions will be fulfilled if not only the body as a whole, but also all of the coloured points as well, coincide with each other after superposition. Such a kind of symmetry is called *coloured*.

Let us consider firstly some kinds of a geometrical symmetry. Obviously, a square can be superposed with itself by rotation around its centre through four different angles: 90°, 180°, 270° and 360°; an equilateral triangle behaves the same by rotation through three angles: 120°, 240° and 360°. On the other hand, any other kind of triangle has only one superposition angle (360°).

The set of movements of a given kind which retains, according to chosen criteria (either geometrical or coloured), a body's identity is called a symmetry group, while the number of such movements is a symmetry order. Consequently, the order of the *rotational* geometrical symmetry for a square is 4, for an equilateral triangle it is 3, and for a scalean triangle it is 1.

All rectangles, as well as isosceles triangles also possess a *reflection*, or *mirror*, symmetry, symbolised by the letter *m*. Bodies with a reflection symmetry have one or more reflection, or symmetry planes. Obviously, a square has four such planes, a rectangle two planes, while an equilateral triangle three planes. It is easy to see that the number of symmetry planes (if they exist at all) is equal to the order of the rotational symmetry. Thus for describing an entire symmetry group of a body one should indicate the number of rotations and the presence or absence of symmetry planes. For example, an entire rotation-reflection group for a square will be *4·m*, for a rectangle *2·m*, for an equilateral triangle *3·m* etc.

Of special interest for biologists are bodies which lack reflection symmetry (while the rotational symmetry can be retained). These bodies show a chirality, or enantiomorphism, and can be arbitrarily divided into left and right forms (Fig. 1.1).

Returning to the bodies possessing reflection symmetry, let us mention those with an indefinitely large order. Such is a circle, or a disc, which can be superposed on itself after being rotated to any angle around its centre. Its symmetry order is described by the symbol $\infty \cdot m$. The symmetry of a sphere (this is the most symmetrical body of all) is described by a symbol $\infty / \infty \cdot m$, what means that it includes an indefinitely large number (a bundle) of

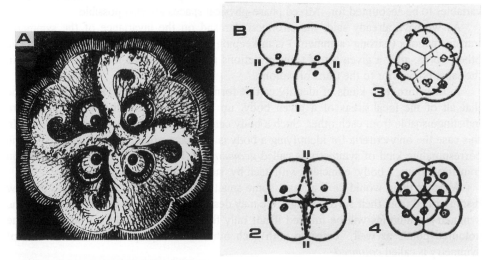

Figure 1.1. Examples of enantiomorphism (lack of the reflection symmetry) in living beings. A: adult scyphomedusa, *Aurelia aurita* (oral view) showing a chirality (enantiomorphism), combined with the 4th order rotational symmetry. B, 1-4: successive frames of enantiomorphic early cleavage in the eggs of a mollusc, *Lymnaea stagnalis*. I and II are the first and second cleavage furrows. At a certain developmental stage (4) the symmetry of a 4·*m* order is temporarily restored. A is from Shubnikov and Koptzik (1972), B is from Mescheryakov (1991), with the authors' permissions.

rotational axes crossing each other in the centre of a sphere under any provisional angles; each of these axes also belongs to a symmetry plane. Bodies with indefinite symmetry order may differ from each other by their symmetry *power*. For example, the symmetry power of a sphere is greater than that of a disk.

The *translational* symmetry is measured by the distance of a shift which is required for superposing a body on itself. A translational symmetry of a homogeneous band without any design on it is of an indefinitely great order, whereas that of a band with a repeated design (ornament) is assumed to be equal to the design's step. The symmetry order of a band with a design consisting of non-repeated elements is *1*.

The introduction of a coloured symmetry immediately reduces the symmetry order of a given body. For example, by putting a coloured mark (or that specified in some other way) onto a surface of a sphere we reduce its symmetry order from $\infty / \infty \cdot m$ to $\infty \cdot m$; a second arbitrarily put mark will reduce it to *1·m*, and the third one will reduce it up to *1* (Fig. 1.2). The increase in the symmetry order is called symmetrization, while its decrease a *dissymmetrization*, or a symmetry break.

As mentioned before, instead of a 'strong' symmetry, demanding the complete identity of a body prior to and after a certain set of transformations we may use as a criterion of a 'weak' symmetry either deterministic or stochastic preservation of some of the body's properties. Let us suggest, for example, that after undergoing some set of movements (rotations through certain angles, shifts of the parts to certain distances, etc.) the physical properties of a body (say, its electrical or optical conductivity) as detected by an immobile

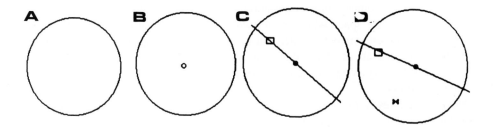

Figure 1.2 A-D. Successive steps in the reduction of a 'point' (colour) symmetry order of a sphere as a result of its surface labelling by an increased number of points of a different quality ($\square$, o and ×). Introduction of three such points (D) reduces the symmetry order of a sphere up to 1.

external observer, remain the same in 100% of cases, or with some lower probability, while all the other movements will change them. That means that the first set of movements creates a symmetry group of the given body in respect to its physical properties. Similarly, *if some set of the mutual shifts of the parts of a developing embryo do not affect the final result of its development, we may say that these shifts* (either caused experimentally or registered within undisturbed development) *comprise its symmetry group according to the developmental criteria.*

Just a few remarks about a temporal symmetry proper, that is 1-dimensional symmetry along a temporal axis. It depends, first of all, upon whether or not we accept the idea of time reversibility. If yes, the group of a temporal symmetry will include the reflection of the future to the past, or vice versa. Hence, we shall have a symmetry $1·m$, plus the translational symmetry of an indefinite order. If, however, we believe in the arrow of time, the reflectional symmetry will disappear and only the translational one can remain. A presence of an indefinite order translational symmetry of time means, however, that nothing really new is happening as time proceeds: a Bergsonian time (firmly associated with a creative activity) should not have such a symmetry!

Meanwhile, the symmetry breaks in Bergsonian time may be also different. If some 'new' X is increased at a constant rate we may speak about an infinitely great translational symmetry of its time derivative (dX/dt = const). Even more typical for dynamical systems, as we hope later on to demonstrate, will be the existence of temporal oscillations which break a pre-existent translational symmetry. This brings into being new *characteristic times*, the oscillation periods. Or, one may speak in this respect about a *finite memory* of a system, a notion to be discussed later on in great detail. In any case, such a profound (and often hierarchical) dissymmetrization of time is intimately linked with the processes of self-organisation.

Let us comment now on the idea of a symmetry in a phase space. Let us suggest that we have a large collection of comparable units which belong to a certain common multiplicity (to a population, in biological terms), but which differ from each other by the values of their properties X and Y. By plotting these properties along the Cartesian co-ordinates we obtain their phase space. Within such a space a given collection will be represented by a *distribution cloud*, which may have different local densities. Let us define as the symmetry movements those translations or 'transplantations' of some parts of this cloud within the phase space which retain (with certain admitted deviations) its identity. Obviously, the highest symmetry order will be shown by a completely homogeneous and indefinitely extended cloud. Any inhomogeneities of a cloud, its contraction towards a smaller area, flattening along a certain axis, or splitting into several subclouds, will reduce the symmetry order.

Now we are ready to describe the main procedure which is necessary for defining unambiguously a process of structurization. It is closely associated with the change of symmetry order.

More than 100 years ago the French physicist Pierre Curie claimed: "C'est la dissymmetrie qui crée l'event" (It is dissymmetry which creates an event) (Curie, 1894). This laconic expression means that there are just the elements of a dissymmetry which permit us to distinguish a body from a surrounding environment, providing it with the elements of individuality: if a body's properties were no less symmetrical than those of the environment, such a discerning could not take place – a body would be invisible.

Accordingly, *we may consider a reduction in symmetry order within a given system as a criterion of its structurization.* The best known, classical examples of such a structurization are so called phase transitions of the first and second order. The processes of a crystallisation (formation of ice from a water) belong to the first order transitions (associated, while moving towards a more structured state, with a substantial heat production), while the second order transitions are exemplified by the appearance of a magnetic moment, or of superconductivity after lowering the temperature below a certain threshold. More detailed distinctions between both kinds of phase transitions are not of a great importance for us. What we would like to stress just now is that all of the phase transitions towards a more structured state are associated with the decrease of a system's symmetry order according to the criteria of its physical properties. Actually, above the phase transitions points the physical properties (optical, magnetic and other) are isotropic, while below the transition points this is not the case. That means, for example, that a rotation of the elements of such a system (according to an external immobile observer) above the transition point will not change at all, or, at least, in a regular manner, its physical properties, while the same procedure performed below the transition point will do that (a crystal possesses certain optical, electrical, etc. anisotropy, while a classical liquid does not). The decrease or increase in the symmetry order is associated with the similar changes of a system's entropy.

The fundamental question is now, whether a dissymmetrization (a creation of an event) may be considered as a spontaneous process, or whether it requires a definite 'cause'. In the framework of a uniform determinism only the second answer is permitted. It was also Pierre Curie who formalised it with great precision in the language of a symmetry theory. The well known Curie principle claims: *"When the given causes generate the given consequences, the elements of symmetry of the causes should manifest themselves in their consequences. If the events show a certain dissymmetry, the same dissymmetry should be revealed in their causes "*(Curie, 1894).

In the other words, the Curie principle forbids a spontaneous dissymmetrization and claims that an acquired dissymmetry should actually pre-exist in a certain (perhaps invisible) form either inside a system which undergoes dissymmetrization, or outside it. Where could it be pre-existent in the case of phase transitions? The reply will be: it is located within the body's particles, let them be the molecules together with their electronic clouds, as during crystallisation, or electron spins, as during para-ferromagnetic transitions. It is the excess kinetic energy which prevents these particles from interacting with each other above the transition point temperatures; these interactions come into play only below the transition point, leading to a mostly dense (of a minimal free energy) particles arrangement. In such a way a 'hidden' dissymmetry of the separate particles comes to a macrolevel, i.e., becomes visible. This elementaristic interpretation of a dissymmetrization warrants both the abovementioned micro-reductionist approach and the Curie principle. Along with that, the Curie principle certainly permits the induction of a dissymmetry in a given body by an external macroscopic agent, if the latter is to the required extent dissymmetrical.

Meanwhile, is the Curie principle really universal? Can we always point to a local dissymmetrizing agent, either embedded within a microstructure of a body, or located outside? Even for the rigid kingdom of the crystalline structures the reply is not so easy. Look to the snowflakes at a high magnification (Fig. 1.3A) or to the ice patterns on window glasses, or to some biological structures of a so-called fractal nature (Fig. 1.3B), to be later on described in more details. In no way can their patterns and symmetry

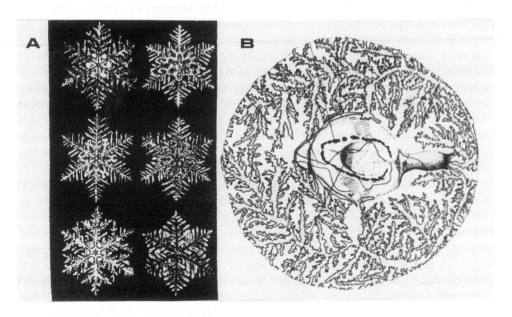

Figure 1.3. Examples of dendritic (A) and fractal (B) structures with their growth rates depending upon the local curvatures of the growing points. A: snowflakes, B: gill patterns in *Planctosphaera* (Hemichordata). A: from Romanovsky et al., 1984; B: from Ivanova-Kazas, 1978 (with the author's permission).

orders be completely derived from those of the constituent molecules or, in the case of ice figures, from any microstructures of the window pane, etc. A theory of these so ubiquitous figures is rather complicated and not yet fully elaborated, but one of its main ideas is that these structures' growth is quite sensitive to the local curvature of a just crystallised part: the greater is its curvature (that is, the smaller is the curvature radius), the greater is the deposition rate of a new material (see, e.g., Langer, 1989). That explains the domination of the branching patterns in these cases. But a curvature, even if quite a local one, is a *collective*, rather than elementarist property: in order to create a curvature an extensive number of molecules should be properly arranged. Meanwhile, the involvement of the collective properties, even if of quite a simple kind, in a shaping process endows the latter with some unusual properties: a high sensitivity to small perturbations and, at least, some elements of memory. For example, the pattern of a growing ice shape largely depends upon its as yet achieved configuration. A growth process becomes under these conditions unique and irreversible, largely contradicting thus the biases of a classical uniform determinism, including Curie principle.

From the crystal-like structures we may pass on to a biologically important class of so-called 'minimal energy figures', consisting of elastic shells or of liquid drops with a considerable surface tension. While free of external forces, these bodies tend to minimise their surface/volume ratio and hence their free energy. At first glance such a process should lead to no more than a sphericalization of a body. Again, however, the situation is not so unambiguous. As far as a tendency to minimise the surface/volume ratio is also (as in the previous examples) a collective, rather than elementarist, process the same properties of a high sensitivity to the initial conditions and to small perturbations and the existence of some memory come into play here as well. While the shapes not so far removed by external forces from the spherical ones when being liberated from the forces immediately approach a sphere, this is not so for considerably elongated tubes (which will be split into a series of drops, thus diminishing their order of translational symmetry), flattened balloons (transformed into toruses), etc. In each of these cases some local minimum of energy is reached, which, however, does not coincide with an absolute minimum. As a result, a geometrical repertoire of minimal energy figures becomes much more attainable than could be expected, and is in no way reducible to their molecular structure and Curie's principle.

In our subsequent account we shall many times refer to the similar kinds of structures, describing them sometimes as 'stress relieving structures'. And now we would like to address ourselves to an even more fascinating class of events, exhibiting in a most pronounced manner the properties of collectivity, sensitivity to small perturbations and memory. What we refer to are so-called dissipative structures, studied by Prigogine's school (Prigogine, 1980; Nicolis and Prigogine, 1977). Meanwhile, one of the first examples of these structures was described quite long ago by the French physicist Bénard.

Bénard observed that some viscous fluids, when poured into shallow plates and intensely heated from below, after some period of time became subdivided into small hexagonal cells which were maintained as long as the intense heating continued (Fig. 1.4A), disappearing immediately after its cessation. The borders of these cells were found to be formed by convective upwards and downwards streams of the liquid particles (Fig. 1.4 B,C). Prepared under some special conditions, the Bénard structures sometimes

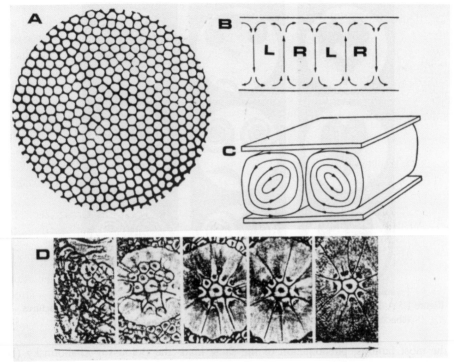

Figure 1.4. The Bénard structures. A: an overall top view. B, C: schemes of the coherent convectional streams with alternated right (R) and left (L) handedness. D: the development of a solitary biomorphic 'autostructure' from a flat layer resulting from a thermocapillary convection. A– C from Romanovsky *et al.*, 1984. D from Michailov and Loskutov, 1990 (with the author's permission).

strikingly resemble the cells of the living tissues (Fig. 1.4D). Generally, the formation of Bénard-like structures is also a kind of a phase transition, but contrary to the classical ones, the structurization is now taking place while passing *over*, rather than *below* a certain temperature threshold: after exceeding a certain temperature, the previously randomly located convective microstreams became collected into macroscopic domains, corresponding to the cells' borders. As a result, an initial homogeneous (and hence highly symmetrical) state of a liquid becomes transformed into a less symmetrical one: while a liquid in its initial state would not change its physical properties under any imaginary shifts and rotations of its parts, that segregated into 'Bénard cells' considerably reduces its order of translational or rotational symmetry and acquires the entities of a new macroscopic scale, in no way pre-existent in the initial medium.

Where does this newly acquired dissymmetry and scale come from? There are no localised 'causes', either micro- or microscopic, which could directly impose them. Now, a dissymmetry has arisen 'spontaneously' or, more correctly, from collective interactions of a great number of particles, each of them separately lacking such a dissymmetry. This view became widely accepted after the discovery and theoretical interpretation of another,

Figure 1.5 A–I. The successive steps in the 'development' of Belousov–Zhabotinsky structures (concentrational waves) within a reaction tube. From Romanovsky et al., 1984.

and the most famous, phenomenon of the same kind: the Beloussov-Zhabotinsky (BZ) chemical reaction (see Prigogine, 1980). In this case a complicated set of the gradually evolved macrostructures (areas of increased concentration of one of the reagents, a bromic acid) arose from an initially homogeneous chemical mixture (Fig. 1.5). These evolving areas of an increased concentration were called autowaves. *"Autowaves exemplify a new type of dynamical processes generating a macroscopic linear scale owed to the local interactions, each of the latter possessing no linear scale at all"* (Krinsky, Zhabotinsky, 1981). The acquiring of a non-pre-existent linear scale is here identical to the reduction in symmetry order.

The term 'dissipative structures' , as applied to Bénard cells and the BZ reaction, means that these dynamical structures, for being maintained, require a continuous and intense enough energy (and in many cases also matter) *in*flow, the incoming energy being partially dissipated. In the other words, they are far from a thermodynamical equilibrium. This is opposite to the above considered equilibrium crystal-like structures born from phase transitions: these demand, instead, an energy *out*flow. As applied to dissipative structures, Curie's principle does not work, or, in more accurate terms, becomes non-constructive: reductions in the symmetry order occurs here spontaneously, or, if speaking more correctly, they come from negligibly small and rare fluctuations which are always present, but in the equilibrium conditions do not leave any traces.

Speaking in more general terms, the existence of dissipative structures violates the principles of a classical uniform determinism, since in these cases the new macroscopic events are not necessarily preceded by any comparable macroscopic causes. An analysis of these phenomena gave rise to new concepts of causality, creating an elegant epistemological system, which we shall consider in the section 1.4. Remarkably, it was elaborated in its general outlines in a purely speculative way by several great mathematicians (Poincaré and Lyapunov being the first) already at the eve of this century, very much previewing its main empirical arguments. Among those who realised that these mathematical discourses, known previously to a few specialists only, provide an adequate language for a new outlook on the entirety of nature, was Ilya Prigogine. He defined a revolutionary change, associated with the discovery and the theoretical interpretation of dissipative structures, as a transition from the classical 'physics of being' to a new 'physics of becoming'. While the first one claims that any event which we naively accept as a new one should necessarily have a comparable individual precursor, bearing its dissymmetry, and so on, up to the infinite past, the 'physics of becoming' accepts overtly the arising of actual new events having no individual precursors.

1.3. Structurization (morphogenesis) in developing organisms.

1.3.1. EMERGENCE OF THE STATICAL AND DYNAMICAL APPROACHES.

What, now, about structurization in the development of organisms? And does it exist at all? This question may sound an absurd one, because the appearance of new structures within development can be perceived in some cases even with the naked eye. It is of interest in this respect, that three hundred years ago the use of the first microscopes inclined the observers to take the opposite point of view (the more so in that it perfectly coincided with the ideas of a uniform determinism): that all the structures observed in advanced embryos and in adults are actually pre-existent (preformed) as very much diminished, but nevertheless individual precursors from the 'very beginning' of development. This idea, being supported by a dominating ideology, appeared to be astonishingly viable, forcing the empiricists to exaggerate the significance of some correct observations (for example, of so called paedomorphosis, or larval propagation in insects), and to ignore or wrongly interpret others. Heroic efforts of Caspar Friedrich Wolff (1734–1794) were needed for proving such obvious things that the gut or the central nervous system of a hen's embryo had initially the shape of a flat plate, rather than a tube, into which these rudiments are transformed later on. We can say that Wolff was the first to notice the changes in the visible symmetry within the development of organisms. These studies should be considered as the starting point of the science of biological *morphogenesis*, that is about the formation of new shapes and structures in embryonic development.

About one hundred years after Wolff, meanwhile, the preformist views were revived, this time in a much more refined form. The reason was that although nobody in those times could by then deny the *de novo* appearance of the *visible* embryonic structures, the possibility could not be rejected that they are preceded by some *hidden* structures, or determinants, arising at a somewhat earlier stage and determining unambiguously, in one to one manner, the formation of the visible ones. Such was the central idea of the so called neopreformism, formulated in the second half of the 19th century by Roux, His, and some others.

Whether the neopreformism was true or wrong, the formulation of such a question marked a very deep shift in embryological thinking which, unfortunately, never attracted adequate attention from an epistemological viewpoint and has as yet not been properly interpreted.

The matter is that before this ideological shift took place each successive stage of development had been regarded as a static, momentarily fixed set of structures, so to say, closed in itself; such an approach can be defined as a static one. By formulating a neopreformist standpoint, researchers express their interest in how a given stage affects its future. In the other words, any developmental stage becomes considered as a temporally extended entity, projected both towards its future and its past (the latter being a future for an even earlier stage), that is as 'a mobile border between the future and the past' whose 'adequate description' makes inevitable 'its transition to the next one' (see second and third quotes). By making such a step researchers started dealing for the first time with real *dynamical* structures, instead of the static ones. The corresponding approach may be thus called a dynamical one.

Using a more formal language, we can denote such an approach as that ascribing to any developmental stage a vector which is defined: (1) in time; (2) in a physical space of an embryo, since most of embryonic areas change, in a more or less regular way, their positions during development; (3) in a phase space of certain developmental characters. In the other words, within the framework of a dynamical approach we should deal with multi-dimensional vectors defined in a 4-dimensional physical space–time and also in a phase space of an arbitrary dimensionality. If we restrict our interest to a morphogenesis proper, that is to a pure geometry of development, we omit the phase space and keep ourselves in the 4-dimensional physical space–time.

This language, which might sound strange and extravagant for a professional embryologist, contains no more than the reformulations of some basic text book notions, most of which have been introduced just during his discussion with the neopreformists by one of the few really great thinkers in embryology, Hans Driesch (1867 – 1941). Those are the notions of a prospective significance, prospective potency (potencies) and a developmental fate (or a developmental pathway). To this we would add the notion of a *competence,* coined later on by Conrad Waddington. Let us briefly explicate and reformulate them.

A *prospective* (or a presumptive) *significance* of a given embryonic area is identical to an adult structure, into which this area will be transformed if the normal course of its development will not be in any way disturbed. Thus, it may be identified with the Aristotelian causa finalis.

The *prospective potencies* comprise a set of adult structures which a given area can produce under *any* experimentally modified conditions compatible with a living state. The main idea of neopreformism was that the potencies of an embryonic area are no larger than its significance; that appeared to be generally wrong.

Now, a *developmental fate,* or a *pathway* of an embryonic area means all the totality of its changes from a given moment of development up to an adult state, be it normal or abnormal. This is just what we imply by introducing our multi-dimensional vector.

A grammatical similarity of the term "potencies" with the "potentials" used in physics is in no way misleading (although if hardly deliberate). Similarly to potentials, the developmental potencies can be defined as multi-dimensional scalars (static structures of an adult) in a vector field of the developmental pathways. On their side, the pathways correspond to the differences of the potentials (differences between a given stage and an adult static state) while their time derivatives correspond to the gradients of the potentials.

A notion of *competence* has been suggested for describing the ability of an embryonic tissue to develop in a certain direction if some more or less specific conditions ('causes') are also provided. In the language which is used here we may consider a competence as both a static and a dynamical notion. As a static one it means a possession by a given embryonic area of a certain set of potencies, while in its dynamical aspect it is *the process of acquiring* this set by a given area. In the both cases the competence is defined in the phase space, and not in the physical one. In its static meaning the competence is a set of scalars in the phase space, while in its dynamical meaning it is a set of multi-dimensional vectors in time and in a phase space. In such a meaning it is in some way complementary to a morphogenetic pathway, which we have considered as being defined in time and in physical space.

Our next task will be to show that not only a static (which is obvious) but also the dynamical approach can be effectively formalised in the terms of a symmetry theory, and that the corresponding terminology can be used for describing, in a very much condensed way, the main types of embryological experiments. For doing this, we must take into consideration not only a static, or *visible* symmetry (that of embryonic structures observed at a given moment of time) but also that of the developmental *fates* (vectors in the phase space) of the different embryonic areas. We shall denote this latter kind of symmetry as F-symmetry.

Let us consider first of all the F-symmetry in the physical space. To detect it means to know, for each developmental stage separately, whether some displacements of embryonic material are compatible with the preservation of a normal adult state's structure. Let us take an embryonic stage *em* (Fig. 1.6), make at this stage some displacements and look whether this will affect the structure of an adult state (*ad*). If and only if this will be realized in a one to one manner, that is if the *ad* structures will be shifted from their normal positions in the same way as it was done with their precursors at *em* stage, shall we conclude that the F-symmetry of the both *em* and *ad* stages is of the same order (Fig. 1.6 A). In classical terms this is the same as claiming that at the *em* stage the potencies have been already prelocalized. If, however, this is not the case and a certain set or range of the displacements at least (if not any displacements) performed at *em* stage will not affect the *ad* structure (Fig. 1.6 B), we should conclude that according to this set of displacement a F-symmetry order of *em* stage is greater than that of *ad* structure (in classical terms that means that at the *em* stage the potencies are still not prelocalized). To detail such an analysis we may insert between *em* and *ad* some intermediate stage E_{int}. If now some displacements made at *em* stage will not disturb *ad* structure, while displacements at E_{int} do this, we will claim that it is *em* stage to possess the greatest F-symmetry order. In the reverse case we have to conclude that the F-symmetry order is increased from *em* to E_{int} stage for decreasing later on again.

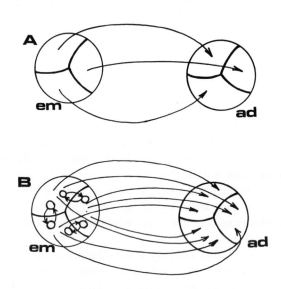

Figure 1.6. The idea of the experiments for testing F-symmetry of embryos according to the displacements (transplantations) criteria. A: both the initial (*em*) and the final (*ad*) stages are of the same symmetry order. B: the symmetry order of the initial stage is greater than that of the final stage because the transplantations shown in B, *em*, by small arrows do not affect the final fates of embryonic parts.

Let us discuss now the notions of a visible and F-symmetry in phase space. What should be stressed first of all is that the phase space symmetries, contrary to those of the physical space, require, for constructing the distribution clouds, a statistical amount of data. This is meanwhile in contrast with the ubiquitous methodology of not only descriptive but also experimental embryology, which well may be based upon single unique observations (guaranteed only by a qualification and a honesty of their author).

At the moment this is only the F-symmetry in the phase space which requires further comments. For estimating F-symmetry we have, first of all, to establish the phase space of developmental fates which we are interested in and plot within this space their distribution cloud for the different developmental stages of the *same* embryonic area (in order not to complicate our task by comparing with each other the areas with different co-ordinates within physical space). The fates may be restricted either to the morphogenetical trajectories, or to the deformations of a given area, or include also the formation of histological structures. In the last case the phase space units might look like the pathways towards skin tissue, muscle tissue, notochord, etc.. In order to determine, for a given stage and embryonic area, the distribution cloud for the selected fates set, it is reasonable to perform at this stage a series of transplantations, or explantations, of this area.

Now we would like to demonstrate that what actually has been done by several generations of both descriptive and experimental embryologists is just studies of the changes in either visible (static) or F- (dynamical) symmetry within the course of development. While most of these studies have been related to physical space, some of them also implied a phase space. A reformulation of their data in symmetry terms would be, in our view, of a great help by condensing and generalising a lot of scattered empirical data.

1.3.2. A VISIBLE (STATIC) SYMMETRY IN THE PHYSICAL SPACE.

What are the main transformations of a visible symmetry during development? Let us briefly, and rather superficially (as a kind of a first approximation) look onto them, taking as an example the early development of one of the most extensively studied organisms – sea urchin embryos (Fig. 1.7).

A fertilized sea urchin egg already has a well defined polarity, marked by so called polar bodies, extruded on the animal pole; the opposite pole is called vegetal. Hence, an egg symmetry may be described by the formula $\infty \cdot m$. Such a polarised egg (Fig. 1.7 A) has been developed from a germ cell possessing initially no preferential polar axis and having thus the greatest possible symmetry order $\infty / \infty \cdot m$.

The first period of development is an egg cleavage, that is, its progressive subdivision into the cells, which are called blastomeres (Fig. 1.7 B-F). In sea urchin eggs this process is characterised by the preservation of a radial (rotational) symmetry in relation to a polar axis, the order of the symmetry being firstly decreased, by a first egg division, from $\infty \cdot m$ to $2 \cdot m$, and then increased, after the second cell division, to $4 \cdot m$, and after the third to $8 \cdot m$. Then the cleavage loses its synchronicity and regularity (Fig. 1.7 G) and we become confronted, for the first time, by a fundamental question: what scale should we use for evaluating the advanced stages symmetry?

If we continue to look upon single cells, the symmetry order of an entire embryo will be immediately reduced up to 1 for never increasing again. Intuitively we feel, however, that this is unreasonable and that for our subsequent estimations we must overlook single cells, paying attention only to the overall contour of an embryo. By doing this we easily observe that at the stage of a so called epithelial blastula (Fig. 1.7 H) the symmetry order is again $\infty \cdot m$, as it was in the fertilised egg prior to cleavage.

As we shall learn in more details from Chapter 2, such a restoration of a symmetry order in no way is a trivial or a passive process: for obtaining a spherical blastula shape, an extensive water pumping inside the primary embryonic cavity, the blastocoel, is required. This is provided meanwhile, by a dissymmetrical arrangement of the ion transporting mechanisms within cells. Consequently, a symmetrization of an entire embryo is associated with a dissymmetrization of its constituent parts, the cells. We shall many times see later on, that such a 'symmetry exchange' between the different levels is something usual.

A brief subsequent period of development – a beginning of a formation of a primary embryonic gut, or a gastrulation (Fig. 1.7 I) – is not accompanied by any symmetry changes, since a gut is initially growing straightway towards the animal pole. Soon, meanwhile, it starts to bend, for approaching later on one of the embryo sides, a ventral body wall. The whole embryo shape also becomes distorted (Fig. 1.7 J). This is associated with the loss of rotational symmetry, while the reflection (mirror) symmetry is still

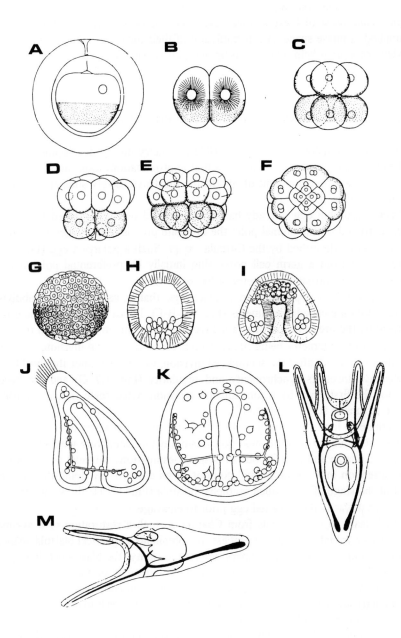

Figure 1.7. Successive developmental stages of a sea urchin embryo. A: an already polarized egg prior to cleavage within its envelopes. B–F: cleavage. G, H: blastula stage; I–K: gastrulation. L, M: a larva, as seen in the frontal and in the saggital projections correspondingly (from Spemann, 1936).

preserved (Fig. 1.7 K); the symmetry order now becomes $1 \cdot m$. The plane of the mirror symmetry is called the saggital plane; it joins the ventral and the dorsal embryo surfaces. Somewhat later the reflection (mirror) symmetry is also broken. This is because one of the lateral outgrowths of a primary gut, which are called coelomic sacs, becomes larger than its partner (Fig. 1.7 L). The largest sac corresponds to the left side of the larval body. In such a way a left–right dissymmetry is established, and the overall symmetry order of the embryo definitely becomes 1.

This is not meanwhile the whole story of the developmental dissymmetrization. For the advanced developmental stages of all embryos there are transformations of a translational symmetry which play a leading role. In sea urchin embryos they are exemplified by a subdivision of the both coelomic sacs into three parts each. Such a subdivision, affecting often an entire body shape, is a very common phenomenon. In its most obvious form it goes on in so called metameric animals, to which most species belong. In almost all the embryos of Vertebrates, for example, an initially smooth neural tube is subdivided, in its anterior part, firstly into three and then into five brain vesicles; an initially more or less homogeneous sheet of the axial mesoderm is gradually split into the metameric units, the somites, and so on.

Thus, the main developmental tendency is in decreasing, step by step the symmetry order of an entire embryo. On the other hand, as we already can see, the opposite tendency is also exhibited from time to time. In some cases (a so called pulsatorial growth in hydroid polypes and other species, to be traced in Chapter 2 in great detail) the ups and downs in a symmetry order are regularly alternated at the same structural level. What is, meanwhile, even most typical, a decrease in the symmetry order of an entire embryo is accompanied by its increase in its constituent parts. These smaller scale symmetrizations may be of quite different kinds. For example, during metamorphosis of a sea urchin larva or its taxonomic relatives one of the coelomic sacs, a so called hydrocoel, is subdivided radially into several equal parts (into five, as a rule), thus acquiring a perfect n-order rotational symmetry. Several kinds of rudiments, among them the sensory rudiments of the Vertebrates are formed from the temporary structures consisting of columnar cells, so called placodes, having within themselves a perfect translational symmetry. Sometimes this symmetry goes down to the single cells level, while in other cases this is not so. Anyway, the existence of an intermediate structural level symmetry, combined with a dissymmetry of the upper and, probably, the lower level structures is an universal phenomenon, quite important for understanding the internal logic of development. Soon we shall discuss the similar events ('fine- and coarse-grain symmetry'), in the way they are related to the developmental fates.

1.3.3. A VISIBLE (STATIC) SYMMETRY IN THE PHASE SPACE

As early as in the year 1828 Carl von Baer, after tracing the development of about 2000 chicken embryos, wrote: "The younger are embryos the greater is the number of distinctions which can be traced, and the greater are these distinctions... One can hardly imagine how such different formations lead to the same results and why, besides normal young ones, one does not observe a lot of monsters. Since, however, the number of such monsters among more advanced embryos and the hatched chickens is very small, one

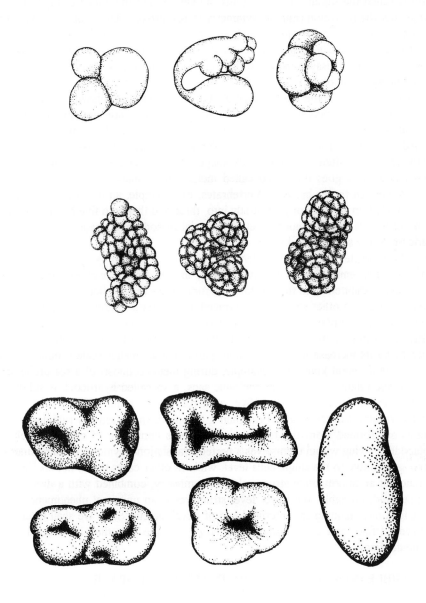

Figure 1.8. The morphogenetic variability in the development of a hydroid polyp, *Dynamena pumila*. Upper row: early cleavage. Second row: late cleavage. Lower part: various abortive "gastrulation attempts". The equifinality of all the different developmental pathways is illustrated by quite a uniform view of a completed larva (lowest row, right frame). Courtesy of Julia Krauss.

should conclude that the differences described are smoothed, and any deviations are brought back to a norm as far as possible". As a result, he makes a remarkable conclusion: "That makes it clear, that it is not any stage by itself which, owing to its own properties, determines the next, but instead, more general and higher relations regulate all of this... it is not the matter, but the essence (idea, according to the new school) of a form of a breeding animal which governs the development of a foetus".

If presenting Baer's data in the form of distribution clouds of the observed characters, the clouds will obviously look more extended and probably homogeneous the younger the developmental stage. For the adults the cloud will be diminished either down to a single point or, most probably, to a correlation line (since the uniformity of the adults is based much more upon the perfect correlation between the different characters rather than their constancy). In any case, the symmetry order of the cloud will be, according to the criteria formulated above, very much reduced with age.

In no way are von Baer's data unique or out of time. Adequately detailed text books on descriptive embryology (mostly invertebrate) are full of data indicating that the adult or the larval structure is much more constant than the embryonic pathways which lead to it. Cherdantsev and Scobeyeva (1994) traced a substantial variability of some dynamic components of amphibians gastrulation, in spite of the entire process looking if it is very precise. A lot of examples of a so called equifinality, that is, of reaching the same developmental results by quite different ways, is shown by hydroid polyps, the representatives of a Cnidaria phylum. Most of them have a larva of a planula type, which is quite uniform throughout the entire phylum. However, the pathways which lead towards this stage are astonishingly different even if tracing the development of eggs produced by the same individuum (Fig. 1.8).

Another, and no less important, aspect of equifinality can be visualised if comparing with each other the characters belonging to the different structural levels: as a rule, those related to higher levels are much more constant and precise than those of the lower levels. Just a small part of the corresponding examples is shown in Fig. 1.9.

How can be such events interpreted in the symmetry terms? If one trace the development of a single organism in a physical space, the equifinality can be well qualified as an increase in the symmetry order. For example, a symmetry order of a planula larva is often $1 \cdot m$, that is largely exceeding the symmetry order of the embryonic stages shown in Fig. 1.8. This is not, meanwhile, a general case: the symmetry order of the whole may well not exceed 1, this in no way violating equifinality. The essence of this phenomenon is not so much in pursuing greater 'shape precision' as in restricting the degrees of freedom of a developing organism. Obviously such an outcome should be considered as a decrease (rather than increase) of the symmetry order, as viewed within a phase space.

1.3.4. DYNAMIC OR F-SYMMETRY IN THE PHYSICAL SPACE.

The very origination of experimental embryology has been associated with the F-symmetry tests. A first really correct and the most famous experiment of this kind (after a partially misleading attempt by W.Roux to injure one of the two first blastomers of a frog's egg) was performed by Driesch more than century ago (in 1892). It consisted in isolating one of the two first blastomeres of a sea urchin egg from its partner; finally, the isolated

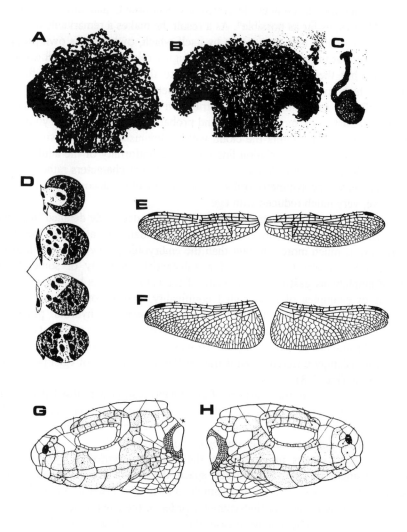

Figure 1.9. Several classical examples of a greater precision of the higher levels structures, as compared with the lower levels ones. A, B: successive stages of the formation of a fruit body in a mushroom, *Marasmius*, by an irregular interplexing of its hyphae. C is low magnification view of the same sample. D: several cross-sections of a chimaeric notochord, produced by the fused tissues of a *Triton taeniatus* and *Triton cristatus* embryos: an overall shape of the rudiment is much more precise than the arrangement of its tissue components. E, F: a variability in the veins arrangement in the left and right anterior (E) or posterior (F) wings of a dragon fly, *Sympetrum flaveolum*, in no way affects the precise shapes of the overall wings. G, H: a similar situation for the skull structures of a lizard, *Lacerta agilis*. Inspite of a considerable variability in the arrangement of the bone plates, the shapes of the left and right parts are exactly the same. A–C from Gurwitsch, 1930; D from Spemann, 1936; E–H from Zacharov, 1987, with the authors' permissions

blastomere gave rise to a complete and proportional, although if diminished, larva. It is of primary importance to follow the path of this restoration process, defined by Driesch as embryonic regulation. After its isolation, a sole blastomere cleaves as normally, thus producing a half-blastula with a large hemispherical opening. Later on, the half-blastula is spontaneously closed into an almost spherical diminished complete blastula. This step is crucial for subsequent regulation to occur: if a sphericalisation does not take place at all, or is largely abnormal (with extensive folds or overlaps), the regulation fails (Beloussov and Bogdanovsky, 1980).

One of Driesch' successors, Sven Hörstadius, has shown by labelling embryonic tissues that during a half-blastula closure its initial vegetal and animal poles are coming into direct contact with each other (Hörstadius, 1939). Consequently the closure goes mostly around the axis perpendicular to the animo-vegetal one (Fig. 1.10, C).

It is easy to see that during such a sphericalisation of a half-blastula each of its previously labelled material points (or areas) except one (which may coincide with the vegetal pole) changes its initial angular coordinates according, say, to the animal–vegetal

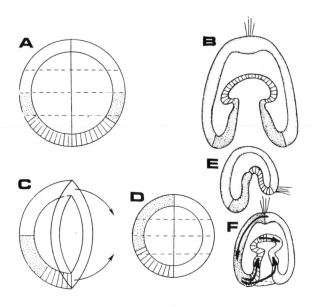

Figure 1.10. Embryonic regulations in sea urchin embryos. A: a fate map of a normal blastula. The presumptive mesoderm is dotted, the presumptive endoderm is hatched and the presumptive ectoderm is left empty. B shows the positions of these areas in the normal gastrula stage embryo. The animal pole is marked by a ciliary tuft. C: a closure of a dissected half-blastula. D: the arrangement of the presumptive areas in a closed half-blastula is very much non-homologous to that in the normal blastula. If the further development of a partial embryo were going according to the presumptive fates of its areas, nothing except a monster like that shown in E could be developed at all. The actually observed regulative morphogenesis demand shifts in the presumptive fates as shown by in F by curved arrows.

axis (Fig. 1.10, cf A and D). In the case of a one to one correspondence between the areas' potencies and their prospective significances, a monster like that shown in Fig. 1.10E should be produced. Since, instead, many of the embryos so produced are quite proportional and geometrically similar to the normal ones (Fig. 1.10, cf F and B), we have to conclude that such a correspondence do not take place at all and that the potencies of all the embryonic areas have also been shifted according to their new positions (Fig. 1.10, F). Let us now take into consideration that the same complete and proportionally structured embryo can be obtained from any isolated meridional sector of a cleaving egg and of a blastula, from eggs with displaced blastomeres, from the fused meridional and vegetal halves of a cleaving egg, etc (see, e.g., Spemann, 1936). That means that up to the blastula stage at least (and in some experiments even up to the gastrula stage) any rotational or translational displacements of the embryonic material are compatible with the formation of a normal larva. Driesch formulated this result by claiming that all the territory of the given stage embryos is equipotential, and all of their cells are totipotent (each cell can contribute to any rudiment). It would be probably more accurate to conclude, that all of the embryonic potencies at a given stage are still *delocalized*. And in terms of the symmetry theory the same may be expressed by claiming that within a given developmental period an embryo possesses, probably with few exceptions, the greatest possible (spherical) order of its F-symmetry.

Several generations of investigators after Driesch continued to study the regulatorial capacities of developing organisms. In symmetry terms, their main results can be summarised as follows:

1. Any organism passes through a certain more or less prolonged initial period of development when it has the greatest possible F-symmetry order (that is, all of its parts are equipotential). The duration of this period may be quite different and in some cases (related mostly to lower invertebrates) it extends over the entire life cycle.

2. As a rule, the F-symmetry order is diminished progressively as the development proceeds, but in a few cases its transitional increases can be detected (that is, a set of the potencies can be temporarily increased).

3. In most cases (and in any case within Vertebrates' development) the dissymmetrization according to F-criteria is pretty well correlated with the visible (static) dissymmetrization, only slightly (if at all) preceding the latter.

4. Similarly to the visible symmetry, the F-symmetry exhibits different dynamics at the different structural levels.

To comment upon the latter point let us consider some recent experiments on the transplantations of single cells at the blastula stage of amphibian embryos (Wylie *et al.*, 1987; Snape *et al.*, 1987; Godsave and Slack, 1991) and compare them with some earlier data obtained while transplanting and explanting larger embryonic areas consisting of several dozens cells at least (Holtfreter, 1938a,b). At the early blastula stage of a frog's egg each one of its cells taken separately is practically totipotent: even a blastomere, destined to contribute to a gut, while being transplanted into the presumptive area of an eye, becomes a retina cell (Wylie *et al.*, 1987). This is not so, meanwhile, for the larger areas; the results of their transplantations and explantations are such that if an embryo has been already segregated into domains with different potencies sets (endodermal, ectodermal, etc.). A few hours later, at the late blastula stage, just the same conclusions can be drawn from the single cells transplantations. For describing all of these phenomenae in symmetry terms, it is worthwhile introducing the notions of a fine grain and a coarse grain symmetry; the fine grains will correspond to single cells while the coarse grains will correspond to the larger areas including several dozens cells. Now we may reformulate the above results by claiming that, as development proceeds, *the rate of a coarse grain dissymmetrization exceeds that of a fine grain one*. This is an important generalisation, stressing the holistic character of a developmental dissymmetrization.

5. At any of its structural levels the F-dissymmetrization is going on in a step wise and almost always bifurcation manner: any initially F-symmetrical (equipotential) territory is split into two, each one of them obtaining a narrowed set of potencies, and so on. That leads to the following fundamental conclusion: any potency of a given stage embryo should be projected to the next bifurcation stage, rather than to the adult state directly; in other words, the adult-oriented potencies are in most cases mere fictions. As to the bifurcation points, their number is always limited and they are adjusted to certain crucial moments of a development; correspondingly, as will be shown later on in more details, the evolution of the potencies is to be split into a restricted set of the *finite* (not infinitesimal) multidimensional vectors in the physical and phase space. Another important and non-trivial conclusion from the classical experimental data is that the territories which correspond to the different sets of potencies can overlap each other, so that the overlapped regions contain the largest potencies sets (Holtfreter, 1933).

1.3.5. F-SYMMETRY IN THE PHASE SPACE

Let us start from considering the purely *morphogenetic* potencies, that is, those associated with the capacities of single cell or multicellular formations to pass, within some finite developmental future, along certain trajectories or to deform themselves in a certain manner. Are such potencies strictly predetermined from the very beginning or do they show instead some uncertainties even within a normal, undisturbed development? Recent advances in vital labelling of embryonic cells have shown, indeed, that the limits of such a variability are much greater than could have been expected before. For example (Dale and Slack, 1987), a progeny of the individual blastomeres, labelled at the 32 cell stage of *Xenopus* eggs appeared to be widely and irregularly scattered between non-labelled cells (Fig. 1.11). The same was true for the cells of a compact labelled piece of the axial mesoderm migrating towards a dorsal embryo midline (Keller and Tibbets, 1989).

For passing now from the purely morphogenetic potencies to those directly associated with tissue differentiation we even do not require any new examples. Returning to the experiments with *Xenopus* blastomeres labelling (Dale and Slack, 1987), we can see, that as far as the above mentioned variability of their morphogenetic trajectories in no way disturbs the histological structure of an entire embryo, the 'morphogenetic errors' should be precisely compensated by the changes in cell fates. For providing this the latter should thus be variable and positionally-dependent at the same time. This is supported by the direct observations: for example, a blastomere A2 (Fig. 1.11A) contributes in the equal percent of cases to the lens, retina and otic vesicle and in some cases also to a sucker. Similarly, the blastomere B1 participates most often in the retina development, but sometimes its descendants can be found as far from this as in the heart rudiment. The voluminous contribution of each one blastomere in a given rudiment can be also quite

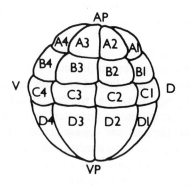

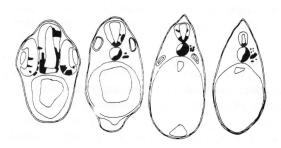

Figure 1.11. A stochasticity of cells' fates in the normal *Xenopus* development, as shown by the individual cells labelling at the 32 blastomeres stage. Upper frame: blastomeres' nomenclature at the 32-cell stage. AP: animal pole, VP: vegetal pole, V: ventral side, D: dorsal side. Lower row: largely disrupted spatial arrangements of B1 blastomere's descendants as seen on the cross sections of the tail bud stage embryos. Similarly to the previous examples (cf Fig.1. 9), the preciseness of the overall morphological structures is much greater than that of the given individual descendants' arrangement . From Dale and Slack (1987), with the publishers' permission.

different. These findings have been confirmed and extended in more recent papers (Vodicka and Gerhart, 1995). However, already the above mentioned classic Holtfreter's (1938 a, b) data led to quite similar conclusions.

Is it possible to extend the notion of a coarse and fine grain F-symmetry to a phase space as well, and if yes, is its dynamics similar to that described for a physical space? In our view, both questions can be answered positively. A 'dimensions' of the grains should now reflect the magnitudes of the potencies sets: coarse grains will correspond to larger sets of 'intermixed ('sticked' to each other) potencies while finer grains will correspond to the smaller sets. The main question will now be, whether a mutual segregation of the larger sets of the mutually sticked potencies will precede the segregation of their smaller sets.

Evidences favouring this view are so numerous that, paradoxically, they are often missed and neglected as belonging to a background. In almost all groups of animals indeed with the possible exception of Sponges and some mosaically developed Nematodes the entire organs and body parts are determined (that is, lose their translational symmetry both in the physical and in phase spaces) earlier and in a much more stable manner than their constituent cells, whose potencies remain to be to a great extent interspersed with each other. Imaginal discs of insects, limbs of vertebrates, and a lot of other rudiments comprise these unique sets of the mutually 'sticked' potencies which, when being observed under higher 'phase resolution' lose to a great extent their uniquieness and specificity and appear to be rather similar to each other. Quite the same phase space F-dynamics is manifested in the earlier developmental stages as well. Returning to the blastula stage of amphibian embryos, we see that practically all the areas of the so called marginal zone (situated between ectoderm and endoderm) share the same, rather large set of mixed potencies. It is a mere fiction to speak, in relation to these developmental stages about the future notochordal, muscular or neural cells; what we have instead are still 'impersonal' marginal zone cells. A segregation of the individual potencies (that is, a diminishing of the 'sticked potencies' grains within a phase space) will take place only later, going hand in hand with the same process in the physical space. From the already mentioned data of Godsave and Slack (1991), the individual cells explanted from a mid-blastula marginal zone exhibit a much wider and substantially overlapped set of potencies than the larger explants from the same areas. We can see that both in the physical and phase spaces a dissymmetrization on a fine grain level is retarded in comparison with that observed on the coarse graines level.

A gradual F-dissymmetrization can be also traced in the realm of single genes expression. It is well known (data taken from Neyfach and Timofejeva, 1978) and even accepted as trivial (although it is quite far from being so), that in sea urchin oocytes about 92% of the structural genes are active, although if exerting quite a low activity, defined as a 'leakage'. Such a situation corresponds to the highest order symmetry in the phase space. Meanwhile, at the blastula stage the active fraction is reduced to about 70%, at the gastrula stage to 37%, and in adults to 10% of structural genes, with a simultaneous increase in the active genes transcription rate.

In any case the existence of a temporal period (and, in many cases, also of a spatial domain) when and where the developmental fates, before becoming finally segregated and restricted, appear to be, so to say, indefinitely mixed with each other, should largely affect our views about the basic principles of development, biasing us towards the self-organisation ideology. Later on we shall return to this point for several times, and especially in Chapter 3. At the moment meanwhile it is enough to conclude that the F-symmetry order in the phase space decreases during the course of development with a dynamics quite similar to other cases of symmetry considered above.

1.3.6. WHAT ABOUT DISSIMMETRIZERS?

So, what about the factors reducing both visible (static) and F-symmetry within development? Is Curie principle really at work in developing embryos? Can we outline the discrete macroscopical dissimmetrizators? Or, perhaps, a macroscopic dissymmetry of an embryo comes from a microdissymmetry of its constituent particles, as during phase transitions of inorganic bodies? Or, at last, will we fail in outlining any discrete macro- or micro-dissymmetrizators and so forced to accept, as in the case of dissipative structures, some kind of collective interactions, making Curie principle non-productive?

It seems reasonable to start from the *macroscopic external dissymmetrizators*. Can we find any macroscopic factors, acting from the external environment, which might be responsible, for example, in establishing the polar axis of an egg, the plane of its mirror symmetry, etc..? Yes, in many cases the response will be positive. The polarity of as yet immature egg of most animal species is assumed to be determined by the asymmetry of its surroundings within the ovary, while that of the freely developed brown algae eggs is oriented by light (a less illuminated egg pole gives rise to a rhizoid, while the opposite pole gives rise to a thallomic plate: see, e.g., Harold, 1990). A saggital (mirror) symmetry plane in an amphibian egg is normally determined by the entrance point of a spermatozoon. A number of other examples of the same kind can be presented. So it looks as if Curie principle is perfectly fulfilled within the earliest developmental stages at least.

This is what is usually written in most of the embryological text books. Not so many specialists know, meanwhile, that the real situation is not so unambiguous. A brown algae eggs become polarized and develop quite normally even if they "are isolated from other cells and from diffusion barriers, kept in dark, and exposed to a gravitational field of only $1g$ – a vector which does not polarize them" (Jaffe, 1968). Similarly, an amphibian egg acquires a saggital symmetry plane even if a spermatozoon is inserted accurately into the animal pole where it cannot act as a dissymmetrizator (Nieuwkoop, 1977), or when it is not presented at all and an egg develops parthenogenetically. And it is absolutely hopeless to find an external dissymmetrizator for each one of the numerous translational symmetry breaks which take place in further development (for example, to find an external blueprint for each one mesodermal somite).

Now what about the *internal microscopic dissymmetrizators*? Is it possible that at least some of the macroscopic dissymmetry of a developing embryo is originated from a microdissymmetry of its constituent parts (molecules, supramolecular structures, cells), similarly to the phase transitions in inorganic world? Again, the situation is ambiguous.

A very special and still mysterious situation is exemplified by a regular (not occasional!) handedness (left–right dissymmetry) of organisms. It seems to be no other possibility of interpreting it than to accept some kind of direct translation of a molecular dissymmetry (most probably, that of the elements of the so called cytoskeleton, the mictotubules) towards a macroscopic level. It is quite difficult to understand, meanwhile, how such an enormous jump of up to no less than 3–4 decimal dimension orders can be possible. One should add to this that handedness is strictly related to the dorso–ventral dissymmetry, which is beyond any doubt established at the macroscopic level, often (but not always) as a result of the action of the external dissymetrizators. In any case, however, the regular handedness is probably the only kind of symmetry for which the elementarist (molecular) origin cannot be rejected.

Let us now discuss the possibilities of a microscopic origin of other kinds of symmetry. The problem of a translation of a microscopic dissymmetry towards the macrolevel can be adequately illustrated by the beautiful studies of Elsdale (1972).

The author traced the   processes of structurization in human fibroblast cultures with and without collagenase – an enzyme which digests collagen fibres. In the presence of collagenase (that is, in more or less abnormal conditions) the randomly seeded fibroblasts gradually form numerous arrays of the uniformly oriented cells (Fig. 1.12, upper frame). This pattern is unstable and "undergoes a slow reorganisation with time... the number of frontiers is gradually reduced, and adjacent cell groups merge as their cells approximate to the same orientation. The end of the process is the transformation of the patchwork into a single uniform parallel array embracing the whole culture and the concomitant disappearance of all frontiers. The global parallel array is entirely stable." (Elsdale, 1972).

The pattern obtained is obviously associated with a dissymmetrization: the cells with an initially random (that is, statistically isotropic) orientation finally acquire a linear one, with a symmetry order $2 \cdot m$. A symmetry of such a kind may be fully derived from that of the single cells; the process of the formation of parallel arrays largely resembles that of a crystallisation or orientation of iron filings in a magnetic field. So far as the adjacent parallel arrays are unstable and smoothly transmit into one another, we may conclude that no new macroscopic structures at a definite scale appear in this case.

Such a kind of a cell behaviour is in no way unique. As we shall demonstrate in chapter 2, a similar process of parallel alignment of the epithelial cells (a so called contact cell polarisation) plays a primary role in morphogenesis. Many types of cells also readily obey the orientation of an extracellular microenvironment. However, as the next example from the same work will teach us, the result may be far more complicated than a simple one to one blueprint of an external dissymmetry.

In the absence of collagenase, that is in the presence of collagen fibers, instead of the above described parallel arrays the peculiar multilayered cell ridges are formed, most of which exhibit a bilateral (reflection) symmetry and consist of several tiers of orthogonally oriented cells (Fig. 1.12, lower frame). Consequently, along with a cell alignment we are here dealing with the formation of new entirely macroscopic structures with their own, low enough symmetry ($1 \cdot m$) and a new dimensional scale (about $2\text{-}5 \times 10^3$ mkm), in no way pre-existed in the initial culture; the microscopic dissymetry embedded in single cells is not enough for bringing such a result into being. We shall discuss the mechanisms of this and other similar structures' formation in Chapter 2. For the moment it is enough to conclude that while the structures of the first type can be considered as being born by a kind of classical phase transition process (which is probably closer to the second order phase transitions), the collagen-mediated structures of a second kind are associated with

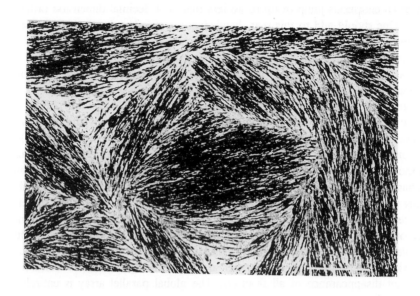

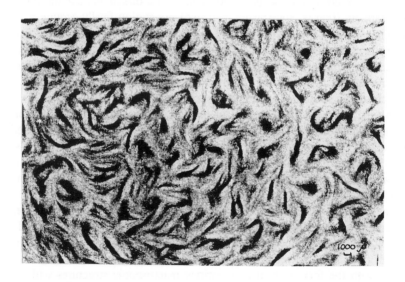

Figure 1.12. Human fibroblasts' patterns in the presence (upper frame) and the absence (lower frame) of a collagenase. For more details see text. From Elsdale (1972), with the publishers' permission.

a *de novo* (by means of collective interactions!) generated macroscopic symmetry. In this respect they definitely approach the dissipative structures. Note that the conditions for the second kind of structures' formation are more natural (collagen deposition non-prevented) than those for the first one.

This is not to say that the developing organisms in the natural conditions do not use at all the microscopic dissymmetrizators; as we shall see in a number of cases, they do that when required, but in the combination of some powerful collective interactions which may, if necessary, smooth out completely the microscopically born dissymmetry. Other fascinating examples of a dissociability of the micro- and macrodissymmetries can be borrowed from Frankel's (1989) studies on the organisation of a so called oral field in Ciliates (Infusoria).

An oral field may be considered as a complicated hierarchical structure assembled by the submicroscopic structures, microtubules, which themselves consist of a glycoprotein, the tubuline. Owing to the tubuline subunits' package, the microtubules have a quite specific chiral dissymmetry, which is reflected in the dissymmetry of the cilia, the derivates of microtubules. Up to a certain dimensional level (roughly up to 1 mkm) the structure and symmetry of the oral field components "is dependent only on the intrinsic properties of the building blocks" (Frankel, 1989, p.143). As to the larger dimensional patterns, meanwhile, the situation changes abruptly: the large ciliary units become oriented within a whole cell body according to an entire body left–right handedness (which may be reoriented in independent ways), rather than according to the chirality sign of the microtubules. These observations show that the microscopically originated dissymmetry becomes non-effective above a certain dimensional limit (which is, as mentioned before, of about 1 mkm). We take a liberty to define this limit as a *Frankel's barrier*.

Among other examples of a possible extension of the single cells' dissymmetry towards an organismic level the problem of the so called disto-proximal polarity of markedly axial organisms (for examples, hydroid polypes) is of interest. Is it possible that the overall disto-proximal polarity of a hydroid stem is directly originated by a superposition of the similar polarities of its constituent cells? There are, indeed, evidences (Labas *et al.*, 1987) that the single stem's cells have a definite disto-proximal polarity, as shown by their electrical properties. Uniform cell orientation makes plausible a direct summation of their polarities in that of an entire stem. However, as shown by regeneration of a new hydranth from a proximal stem piece (Kazakova *et al.*, 1991) or by transplantations of the reversed stem pieces (Kossevitch, 1995) an overall stem's polarity well can be inverted in relation to the single cells' polarities.

In the broader view, it is the variability of the static and dynamical elements of the developing organisms, which we have already traced in great detail, which makes it hardly possible to extend the microscopical polarity towards the macrolevel (with the exception of a left–right regular handedness, as mentioned before). In any case, that makes us to drive our attention to what may be defined as the macroscopic internal dissymmetrizers. In the other words, we will be interested in knowing, whether some definite rudiments of an embryo may serve as either static or (and) dynamical dissymmetrizers, that is, may either directly decrease the symmetry order of an embryonic shape, or (and) restrict the potencies of a given embryonic area. And here we return again to Hans Driesch with his famous fate determining law.

This law, which Driesch considered as a direct deduction from his experiments on embryonic regulations, have been formulated in quite a laconic manner:

"The prospective fate of an embryonic part is a function of its position within a whole" (Driesch, 1921).

Insofar as we are dealing with multipotent systems, the determination of a final fate of an embryonic part means a selection of one potency out of its whole set, that is a typical F-dissymmetrization. That permits us to reformulate Driesch's law in the following manner:

F-dissymmetrization of a given part of embryo is somehow linked with this part's position within a whole.

How can such a linkage be possible? *A priori*, we may outline two following versions:

(1) a fate determining position is referred to some discrete and previously distinguished ('privileged') material elements of an embryo. In this case, for providing dissymmetrization in a 3-dimensional space we need as a minimum 3 such elements (differing from each other by their internal properties); for doing the same in a 2-dimensional space we need 2 such elements; etc..

(2) a fate-determining position is referred in some way to an indivisible whole, irrespective of the mutual positions of any its elements.

What can we derive from embryological experiments? Some of them, consisting, for example, in transplanting fairly small embryonic areas into unusual positions within an undisturbed whole, are compatible with both alternatives. Under these conditions the first type of interpretation is usually preferred, looking more 'understandable' and compatible with the immortal ideology of uniform determinism. The ideas of the 'organisers' and other 'active centres' as bearers of 'positional information', affecting more 'passive' embryonic material belong to just this category.

It is not so difficult to demonstrate, meanwhile, that already the first Driesch' experiment in which he obtained a whole larva out of a half of a cleaving egg or of a blastula stage embryo is compatible only with the second alternative. For proving this let us remember that before commencing the regulation process itself, a half-blastula embryo is rolled into a sphere, most of the rolling going around the meridional axis (see Fig. 1.10 C). Obviously, any previously distinguished ('privileged') material elements of an embryo will occupy, after the rolling, the positions which are geometrically non-homologous to those occupied by the same elements in the intact embryo (Fig. 1.13). In the embryos restored from, say, 1/4 or 1/8 of a cleaved egg, or, as in Hörstadius (1939) experiments, from its meridional half merged with the animal one, any element's position will be different each times. Under these circumstances, if the fate determining positions of the embryonic cells were to be referred to any pre-established ('privileged') elements, no embryonic regulations could take place at all. Since they exist, meanwhile, our mostly cautious conclusion would be that the fate determining positions of all the embryonic cells should be referred to according to the *geometrically homologous points* of either normal or partial embryos, *irrespectively of what material elements (elements of what origin and/or internal properties) occupy these points*. We may however make another step and ask ourselves, why after rejecting a linkage of the reference points with some pre-established specifical the material elements do we still believe that the number of the reference points is so restricted? Why not suggesting, as a kind of a generalisation, that there are *all* the elements of an embryo, irrespective of their origin or specificity which

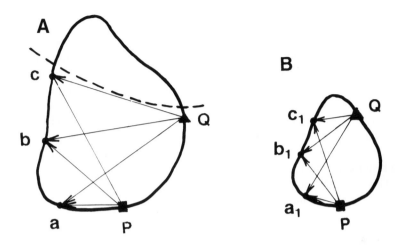

Figure 1.13. Embryonic regulations are incompatible with the assumption of any pre-localized 'privileged' material elements (say, P and Q, frame A), regarded as the sources of a positional information (PI). After the dissection of a part situated upper from the dotted line shown in A and closure of the wound (B) the positions of the elements P and Q (as well as all the others) will become geometrically non-homologous to the same elements' positions in A. As a result, any points of an embryo which occupy in A and B homologous positions (say, a and a_1, b and b_1, c and c_1) will perceive quite different PI signals which is incompatible with embryonic regulations.

equally participate in measuring the fate determining position of all the other embryonic elements? That means that, in the general case, each embryonic element should, on the one hand, perceive the fate-determining impulses and, on the other hand, participate in generating them. In both cases equivalent roles are played by the elements occupying geometrically homologous positions, irrespective of their origin, internal properties, etc.. In such a way we come to the formulation of an essentially collective and holistic mode of establishing the fate determining positions of embryonic elements or, in the other terms, of providing their F-dissymmetrization.

The arguments, based upon the analysis of Driesch's regulations, are far from being the only ones favouring this point of view. The abovementioned outstripping of a fine grain dissymmetrization by a coarse grain one seems to be no less convincing in supporting a holistic and a 'descending' character of the process.

By summarising all the above, we may describe a dissymmetrization strategy of a developing embryo as a rather opportunistic one: when available, and most of all at the earliest stages of development, it readily uses the external dissymmetrizers, thus obeying the Curie principle. The cells can use also the microdissymmetry of an external environment, resembling classical phase transitions. In the absence of discrete dissymmetrizers meanwhile, and, in particular, during embryonic regulations, embryos can decrease both their static (visible) and dynamical (F-) dissymmetry without any definite local dissymmetrizers, now imitating the dissipative structures.

Our final estimation of the relative weights of the different modes of dissymmetrization will depend largely upon which one of them is dominated in the normal development. The question is whether a collective, holistic, and largely delocalized mode   of F-dissymmetrization exhibited by embryonic regulations is restricted by just these artificially induced phenomena, while normal development is to a great extent governed by the local, well-defined dissymmetrizators, or whether the reverse is true? A discussion between these two fundamental concepts will probably never be finished. There are, meanwhile, in our view, enough evidences for taking the second of them as a working hypothesis at least, most of all because it seems to be heuristically promising and fruitful. A set of hypotheses linked with the assumption of a holistic regulation in embryonic development is traditionally associated with the idea of a *morphogenetic field*, firstly formulated by Gurwitsch (1914, 1922) and later on elaborated by the different authors in quite different ways (for a historical review see Haraway, 1976; Gilbert *et al.*, 1996). Meanwhile, by the last (1944) Gurwitsch definition the main function of a field is in 'transforming a part of an excitation energy of the molecules of a living cell into a kinetical energy, directed along a field's vector'. This fits perfectly with the idea of a field working as the main dissymmetrizer of biological events. However highly we might evaluate this insight, our modern views about morphogenetic fields differ from those of Gurwitsch and his contemporaries in at least two principal points:

(1) Instead of believing in an unique, single, and vital field factor we are inclined today to accept the existence of a whole hierarchy of 'fields' belonging to the different space–time scales (levels), so that the specificity of the living beings is largely smoothed throughout the entire hierarchy.

(2) The vectorial and positional effects of the fields should be considered, if we paraphrase Krinsky and Zhabotinsky's statement (see p.16), as the results of the collective interactions of a large number of units, each, while being isolated, lacking both the vectorial properties and any capacities to affect other units in a position dependent manner. Such a view is in a strict opposition to the classical interpretation of the physical fields, implying the existence of their discrete microscopical sources.

These new approaches to the classical problem are ultimately based upon the notions and ideas of a self-organisation theory, up to now only briefly mentioned. It is now the proper time to review it in more details.

1.4. The elements of a self-organisation theory relevant to biological morphogenesis.

From the formal point of view, the theory of self-organisation should be regarded as a branch of mathematics, which is sometimes called by mathematicians (in a more technical manner) the qualitative theory of differential equations. Our approach to it will, however, be rather far from that of the professional matematician. Instead of studying how to solve the different tasks or prove the theorems, we will be interested, first of all, in understanding the system of notions elaborated in this branch of science, and to see how they can be applied to the developmental problems. For me, personally, it was quite amazing to realise that such a system, introduced into mechanics and mathematics several decades ago for solving quite specific problems, appeared to be so appropriate for such a huge set of natural events, including those taking place within living organisms. This looks like a strong argument for the fundamental commonness of the a priori introduced mathematical laws and the natural world.

We shall start our account by discussing briefly the basic 'ideology' of ordinary differential equations, one of the most powerful tools invented by the human mind for exploring Nature. Let us stress firstly that they contain, from the very beginning, a rudiment of a Bergsonian idea of an *active memory* directed towards a future. Taking a differential equation in its most general form:

$$\frac{dx}{dt} = F(x)$$

one can claim that the instantaneous state of a system, x, is endowed by a certain active memory since it determines the system's evolution (dx/dt) during the subsequent infinitesimal time period. True, the main efforts of the inventors of the differential calculus were directed not towards extending a system's memory within the temporal dimension, but to the precise opposite, namely to splitting it into infinitesimal periods. However, the idea of the interaction of an instantaneous event with its future was already present.

This is closely related to the following rarely mentioned point: any differential equation contains the idea of a *feedback* between a scalar x value, exemplifying a momentary state of a system, and the vector of its evolution during the next infinitesimal time interval dt. This dependence can be presented in the form of a vector field. All of this is quite adequate, although not enough for describing the developmental dynamics of organisms.

What will happen if we pass now from the linear (first order) to non-linear (higher orders) differential equations? At first glance, nothing major, except that the feedback will become more pronounced (a small shift in the value of x will generate much greater response). Actually, however, the consequences will be much more important. First of all, as evidenced by an eminent physicist, non-linear equations, in contrast to linear ones, describe the events which cannot be reasonably split into isolated components: "It is clear, that not only is the analysis in terms of spatially separate objects generally irrelevant in the context of such [non-linear] theories, but so also is the notion of analysis into more abstract constituents that are not regarded as separate in space" (Bohm, 1980). For apprehending this statement in its general outlines, let us recall, that in the case of a linear graph it is enough to see a small piece of it for extrapolating a whole graph; but this is not so for any non-linear graphs. In any case, it is highly remarkable that any non-linear equation implies *ex definitio* a property of wholeness.

The most general idea which we would like to demonstrate here is, that the non-linear differential equations describe *multilevel systems, capable of self-complication* (of spontaneous decrease in symmetry order, creation of completely new space–temporal structures, etc.), so that their analysis will teach us much about how similar things are done by the developing embryos.

1.4.1. DYNAMIC AND STRUCTURAL STABILITY/INSTABILITY; PARAMETRIC AND DYNAMICAL REGULATION IN 1-DIMENSIONAL SYSTEMS

General definitions
We shall study these fundamental notions using the simplest example of as yet linear differential equations

$$\frac{dx}{dt} = -Kx + C \qquad\qquad (1a)$$

$$\frac{dx}{dt} = Kx - C \qquad\qquad (1b)$$

Here x is called the dynamical variable, while K and C are the parameters. The parameters and the dynamical variables differ from each other by what is defined as their *characteristic times* (T_{ch}): the T_{ch} -s for the parameters are always in an order or more greater than those for the dynamical variables. That means that the parameters, *ex definitio*, change their values at an order or more slowly than the dynamical variables do (correspondingly, the latter are also called the fast variables). From this moment on we will describe the values obtaining different T_{ch} as those belonging to different levels, the greater being T_{ch} the higher would be the level. The changes in a system's behaviour caused or induced by modifying the parameters' values we shall call a *parametric regulation*, while those caused by modifying the dynamical variables we shall call *dynamical regulation*.

Presenting (1a, 1b) as the graphs (Fig. 1.14 A,B), we will fix, in the both cases, the values of the parameters K and trace the dependence of the derivative, dx/dt, upon the value of x; such a dependence can be imagined in the form of a one-dimensional vector field, projected onto the x-axis. We can see that in each case the field has a singularity (an unique point) at $x = C/K$. For $K < 0$ (Fig. 1.14 B) this is a point of the vectors' convergence, while for $K > 0$ (Fig. 1.14 A) it is a point of their divergence. A system which has a convergence point is called *dynamically stable* (and the point is called a stable nodule), while a system possessing a divergence point is called *dynamically unstable* (as having an unstable nodule). The dynamical stability/instability is a rudiment of the self-organisation, since it disturbs one of the main rules of uniform determinism – proportionality between causes and effects. It is obvious indeed that in unstable systems any deviations of a dynamical variable, however small they may be, are amplified, while in the stable system they are damped. No less obvious is that whether a system will become stable or unstable depends exclusively upon the sign of the K parameter, while the value of x plays no role at all. In other words, the stability/instability property is the object of the parametric, rather than dynamical regulation. The role of the parametric regulation is mostly obvious by presenting the system's 'portrait' in K, x coordinate set (Fig. 1.14C). We can see now clearly that there are parameters values which determine whether a system exists at all (has a stable solution under $K < 0$) or is unstable throughout. This

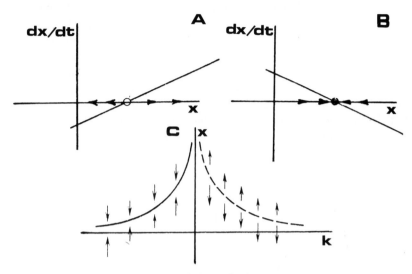

Figure 1.14. Graphs for equation (1). For comments see text.

diagram permits us also to comment upon the important notion of a parametric or *structural* stability/instability: the area in which small changes of parameters values lead to drastic (qualitative) changes in the entire system's structure is unstable, while the area in which the finite perturbations of the parameters do not lead to such changes is structurally stable. In our example the neighbouringhoods of $K=0$ are structurally unstable, since in this area even the small parameters shifts are enough to transform a stable system's state into an unstable one, while the regions outside this neighbourhood are structurally stable. A widely used notion which is similar to the structural stability is the *robustness* of a system (see, for example, Goodwin, 1994). Robustness is regarded as an important and indispensible property of nearly all the natural systems: as claimed several decades ago by the Russian physicist Andronov (see Romanovsky *et al*, 1984), non-robust systems are actually non-observable, since they should take a new appearance each time. For all the living, and in particular for the developing, organisms this property is of a particular importance, since it allows them to keep the individuality (although if not a complete identity!) within an extremely 'noisy' external and internal environment.

This is probably all that we can derive from the analysis of first order differential equations. For enlarging our knowledge of self-organisational capacities, let us now pass directly to third order equations, missing the second order ones.

Bifurcation models
Let us consider the equation

$$\frac{dx}{dt} = Kx - K_1 x^3 \qquad (K_1 > 0) \qquad (2)$$

where $K_1 > 0$. By solving the equation $0 = Kx - K_1x^3$ algebraically we obtain the following roots: $x_1 = 0$, which is real for any value of K, and two other symmetrical roots $x_2, x_3 = \pm\sqrt{K/K_1}$ which are real only when $K > 0$, and which are smoothly deviated from x_1. This is mostly visible if we depict the system in the coordinates K,x (Fig. 1.15A).

Now we evaluate the signs of the derivative dx/dt in the different regions of K,x space separated by the roots. By doing this it becomes obvious that when $K < 0$ the only real root $x_1 = 0$ is stable (it corresponds to the line of the dx/dt vectors convergence), while when $K > 0$ this very solution becomes unstable (the vectors diverge from this line), whereas the two other symmetrical solutions are stable. What we get here is the simplest kind of *bifurcation*, a splitting of one stable solution into two that are also stable, with a simultaneous loss of stability of the initial solution.

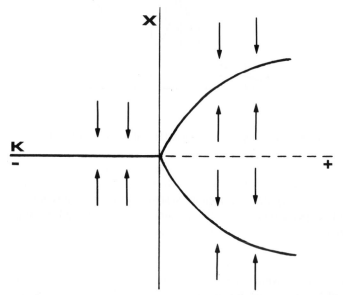

Figure 1.15. A graph for equation (2). Horizontal axis: parameter K, vertical axis: dynamic variable x.

This mathematical construction may be considered as the simplest model of self-complication of a system. Notably, the appearance of two stable solutions instead of one means a creation of two new structures out of a common 'ancestor'. We may also consider such a transformation as a dissymmetrization of a system: when $K<0$ a system has a translational symmetry of an indefinite order (while moving along the K axis) and a reflection symmetry about any vertical plane, whereas when $K>0$ both kinds of symmetry disappear. Now, what about a reflection symmetry about the $x = O$ axis? Here we have an interesting situation: on the one hand, in paying attention only to the entire (parametric) structure of the system, this kind of symmetry will be retained for any value of K. On the other hand, if we want to trace the behaviour of the dynamical variable x immediately to the right of the point $K = 0$, we can see that its symmetrical position

($x = 0$) becomes unstable and hence improbable: under any fluctuations, however small, it will 'jump' to one of the new stable solutions. Thus, in relation to its dynamical structure, the system should lose, for positive K, even this latter element of its symmetry.

This construction has obvious parallels with the fundamental properties of biological development discussed above. Firstly, a set of stable solutions may be identified with the set of potencies (P) set of an embryo. In such a framework a passage from negative to positive K means an increase in the set of potencies and, in other words, a dissymmetrization in P- (or in F-) space (this space may be either physical space, or a phase space, depending upon the meaning of the x variable). Secondly, we can identify the reaching one of the two right half stable solutions by a dynamical variable x with a final selection of one of the two potencies, that is, with the determination of the fate and the establishment of a visible dissymmetry.

What about the factors of dissymmetrization and the obeying of Curie principle? We can see that the situation is no less dubious than in the developing organisms. On one hand, at $K > 0$ any macroscopical shift of the dynamical variable towards positive or negative will irreversibly, and in a completely predictable way, determine the final dissymmetry: in this sense it is a Curie-obedient system. On the other hand, meanwhile, after coming during its rightwards movement along K axis towards the positive K values, the system will, owing its instability, most probably obey a first unpredictable fluctuation, rather than 'wait' for a directional dynamical shift. In other words, the system will most probably reduce its symmetry order in a quite unpredictable way and without any definite dissymmetrizer, just from mere 'noise'. And in any case, the main factor determining the inevitability of the system's dissymmetrization will be the systemic (upper levels) one, definitely associated with the parameters.

Consequently we can see that our system is an object of a dualistic multilevel regulation: firstly the parametric and then the dynamical one. The parametric regulation is itself bilevelled: the upper level is exemplified by a completely constant parameter K_1, which always takes the positive value: otherwise the system would not exist at all. Next down will be the K parameter which actually changes its value, although at an order or more slowly than the dynamical variable does. It is this parameter sign which determines what set of potencies (one or two) the system has. This is something like establishing the constitution of a state, providing or not providing a two party political system. On the other hand, those are the lowest level variables (the dynamical ones) which, acting within this constitution constraints, determine the final choice: if, owing to the fluctuations, the value of x becomes positive, it will be the upper stable solution that is selected, and vice versa. It is worth mentioning that such dynamical regulation is highly degenerative: *all* of the positive x values will bring a variable towards the K-determined upper solution, while all the negative ones will bring it to a lower solution. Such a high degeneration means that the system is very robust in relation to dynamical (fast variables) noise. That contradicts a widespread deterministic view that any precise macroscopic results can be produced only by a no less precise microdynamics. In biological terms this our conclusion perfectly correlates with the above described phenomena of developmental equifinality.

From the most general point of view, the dualistic parametric/dynamical regulation endows a system with a substantial 'internal freedom' which is really amazing if comparing it with the preciseness and reproducibility of the results achieved. What we mean is not just a freedom in selecting each possibility out of two, but, most of all, the already mentioned freedom of the dynamical variable to take no less than semi-infinite number of different values without affecting the overall developmental result. Such a 'hidden freedom', which is, as we can see, an inherent property of a large class of non-linear differential equations and, hence, of all the natural systems obeying them, is an indispensable source of potential novelties, which make it possible the biological and social evolution, even if in danger of certain malformations.

Inspite of all its interest, meanwhile, the system *(2)* has, from a developmental point of view, some serious deficiencies. The main one is its non-reliability: a final state to be achieved is actually thrown upon the mercy of the fluctuations of the dynamical variables, and is hence unpredictable. Is it possible to reach some higher reliability without losing other suitable properties?

In mathematical terms, at least, it is not so difficult to do this: it is enough to introduce a quadratic term into the right part of the equation *(2)*. A new, somehow modified equation will look like

$$\frac{dx}{dt} = -Ax - Bx^2 - Cx^3 \tag{3}$$

By exploring this equation in a manner similar to equation *(2)* we discover that under all of the values of A parameter there exists a rational root $x_1 = 0$, and that at the point of A,x space with the coordinates $A = B^2/4C$ and $x = -B/2C$ two other rational roots, x_2 and x_3 appear which diverge symmetrically to the left of this point, as shown in Fig. 1.16; in all the regions of phase space a root with the intermediate x values is unstable (dynamical variables diverge from it) while both extreme roots are stable.

Let us explore this system qualitatively, suggesting that the parameter A is slowly moving from right to left. We can note several interesting properties. Firstly, as we have already noticed, the roots x_2, x_3 are not smoothly deviating from the first one, as they did in the preceding model, but emerge instead in a new (previously non-specified) point of a phase space. That illustrates the possibility of a sudden appearance of 'new' potencies.

The main advantage of this system in relation to *(2)* is meanwhile the existence of a so-called *metastability zone*, which corresponds to the area $0 < A < B^2/4C$ of the parameter A values. Within such a zone the finite, not too small perturbations of a dynamical variable are required for transforming a possibility to reach a new stable solution (x_3) into a reality. It is a requirement for a *finite* perturbation which provides the system *(3)* with a reliability and a predictability of its behaviour, since such a perturbation can be, so to say, properly arranged by internal regulatorial mechanisms of a developing organism; by using such an opportunity the system will become Curie–obedient. And only in the case of a very smooth (non-perturbed) movement of the A parameter from right to left via the zero point will a system 'miss' such an opportunity, becoming again thrown upon the mercy of the fluctuations, similarly to the system *(2)*.

Because of the existence of a metastability zone, in the system *(3)*, contrary to *(2)*, the moment T_1 of achieving a competence (this is the moment of reaching by the parameter A, in its movement from $+\infty$ towards left the value $B^2/4C$) can be temporally separated from the moment T_2 of the final selection of a given potency from the entire potencies set (this

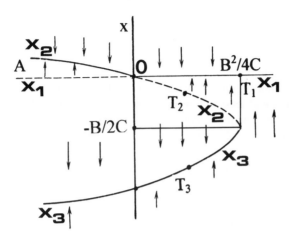

Figure 1.16. A graph for equation (3) in the space of a parameter A (horizontal axis) and a dynamical variable x (vertical axis).

will be the moment, when a dynamical variable will cross, under the influence of a dowanwards directed perturbation, the instability barrier, shown by dotted line). Moreover, T_2 can be also temporally separated from the moment T_3 of achieving the final, visible results of development (this is the moment, when the dynamical variable, moving further down from the instability barrier, reaches its final value at a new stable solution branch). In other words we can now distinguish from each other the time moments which correspond to the stages of establishing a competence for a given differentiation (this is T_1), of a determination (T_2) and of a final, visible differentiation (T_3), as they are defined in classical embryology. Such an ease in reformulating the empirical embryological notions in the strict terms of a self–organization theory seems to be remarkable. We can also see that the slowest, and hence main time–limiting process is that of establishing a competence, since it is, in mathematical terms, a parametrically regulated event. Both determination and differentiation are the dynamical, and hence faster, events. A visible slowness of these processes should be ascribed to the fact that they are actually 'enslaved', following Haken's (1988) terminology, by a slowly evolved competence.

The asymmetry of the x_2, x_3 solutions in the system (3) also has its biological parallels. As it is well known in embryology (see Chapter 2 for further discussion), in most of the real developmental bifurcations one of the selected ways is, so to say, more 'basical', while another is optional. Such an asymmetry is also related to the notions of the senior and junior modes, to be later on explicated.

We have paid so much attention to the bifurcation systems because there are good reasons to consider them, and in particular (3), as the universal elements of a developmental self-complication. To reach a far-reaching complication by a chain of such successive bifurcations is much safer, 'cheaper' and easier than, for example, passing via trifurcation(s). The latter demands much a higher degree of non-linearity (that of the 5th order), which is difficult to achieve, dangerous (owing to too great a feedback) and non-reliable for correctly selecting the potencies (owing to narrowness of the phase space areas between the neighbouring solutions). However, in some specific cases (for example, in the differentiation of antibodies producing cells, B-lymphocytes) such an explosive and poorly regulated process probably has its advantages. In any case, we have already presented in the previous section the biological evidences supporting the bifurcation character of the selection of developmental fates.

How can we obtain a prolonged succession of bifurcations instead of a single one? Mathematically there are at least two different ways of doing this. The first one is to use some fractal generating functions. This will be discussed somewhat later (pp. 54-56). Another is to introduce (while using the same functions *(2)*, *(3)*) some kind of abrupt parametric changes, being themselves functions of a trajectory, passed in K, x space. Let us return, for the sake of a simplicity, to the equation *(2)* and assume that by moving along $+ \sqrt{K/K_1}$ branch and reaching a point with the coordinates $K = + a$ and $x = + b$ the equation *(2)* will be abruptly transformed into

$$\frac{dx}{dt} = (K - A)(x - B) - K_1 (x - B)^3$$

this being identical to the abrupt introduction of new parameters A, B. As a result we shall obtain a new bifurcation at a given coordinates point. By continuing the same procedure we can obtain an entire series of successive bifurcations (Fig. 1.17). Now, our basic idea

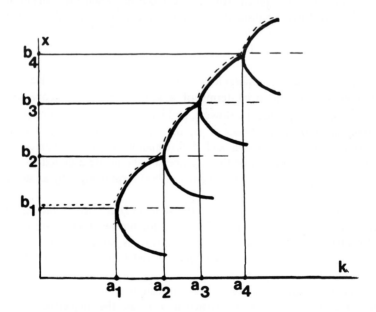

Figure 1.17. A construction of a succession of the bifurcations. For comments see text.

is that the smooth evolution of a system in the parameters/dynamic variables space leads, from time to time, to abrupt changes in some of its other parameters values. The parameters themselves become the objects of the non-linear feedback actions, rather than remaining absolutely independent and constant in their values. This assumption looks as a biologically realistic one, is associated with the notion of a developmental memory and will be exploited many times later on.

The last question in this section will be: what might be the physico-chemical tools for making bifurcations? Is it easy to create them?

Up to now in most cases self-organisational models are interpreted in the terms of a chemical kinetics. In such a viewpoint, the model *(2)* implies a linear autocatalysis (a positive linear feedback between the amount of a substance x and the rate of the reaction dx/dt, exemplified by the first member of the right handside of the equation), combined with third order autoinhibition (the term K_1x^3). The model *(3)* implies three autoinhibiting processes with different coefficients: a linear, a quadratic, and a cubic one. Such a situation may be considered as rather robust and general.

However, the chemokinetical interpretation is far from being the only possible one: a fairly simple mechanical interpretation is also available, in particular in relation to *(2)*. Let us suggest that we have a straight string which is stretched at $K<0$ and compressed from its edges at $K>0$. The parameter K becomes now a measure of stress. While being stretched, a string obviously has only one stable state, namely that of a straight line. Under compression meanwhile this state loses its stability, and infinitesimal deviations from a perfectly coaxial compression (or some microdissymmetries in the string structure) will be enough to bring it quite rapidly (with a rate of a dynamical variable) to an arch shape, which may be bent either up or down with an equal probability. This is the case of so called Eulerian instability, having many applications to morphogenesis (see below). The metastability models, like *(3)*, also may be interpreted in a mechanical way by introducing some finite resistance to bending of the string (Green et al., 1995).

1.4.2. 2-DIMENSIONAL SYSTEMS

The one variable non-linear differential equations considered above are not enough for grasping all of the possibilities of the dynamical systems. Much wider opportunities for self-organisation are provided by non-linear systems including two different dynamical variables (X and Y). We shall review here two examples of such systems: a so called corepression model and a model generating auto-oscillations together with the related regimes. The general methods of qualitative analysis of such systems are briefly as follows.

If we have a two variables system

$$\frac{dx}{dt} = f_1(X,Y)$$

$$\frac{dy}{dt} = f_2(X,Y)$$

which we would like to explore, the first thing which we must do is to solve the ordinary (algebraic) equations

$$0 = f_1(X, Y)$$

$$0 = f_2(X, Y)$$

and to plot the graphs of the solutions on a phase space with the coordinates X, Y. These graphs are called zero-isoclines, since they correspond to zero values of the derivatives. The first isocline ($dx/dt = 0$) can be also denoted as an isocline of the verticals (since the trajectories of the dynamical variables cross it vertically), while the second one ($dy/dt = 0$) is an isocline of the horizontals. The isoclines divide the phase space into areas of qualitatively uniform behaviour. Intersection points of the isoclines may be the points (nodules or focuses) of stability, instability, or stability–instability (so called saddle points). Their stability or instability can be evaluated by determining the signs of dx/dt and dy/dt in the different areas of a phase space.

The model of corepression
This model (see Romanovsky et al., 1984, for more details) has been suggested initially for interpreting a particular case of a corepressive interaction between two genes. It is, however, of a much wider interest.
    The model contains the following equations describing the production of the metabolites X and Y:

$$\frac{dx}{dt} = \frac{A}{1 + Y^2} - X$$

$$\frac{dy}{dt} = \frac{A}{1 + X^2} - Y \tag{4}$$

Exploring these equations as mentioned above, we can see that when $A<2$ both isoclines intersect in one point located on the bisector of the coordinate angle, and this point is a stable nodule (Fig.1.18, a). Meanwhile, when exceeding $A=2$ two other intersection points appear which are now stable nodules, while the initial (now the intermediate) nodule is transformed into a saddle point: the trajectories of the dynamical variable, when approaching the bisector, converge towards this point, but then diverge from it towards one of the newly established peripheral nodules (Fig. 1.18,b).
    This model shows us once again, that the set of potencies of a system (its stable states) is determined by the parameters' values, while the selection of the potencies is determined by the dynamical variables which act, as previously, in a very degenerate way. A new element is the behaviour of the system at $A \cong 2$. Under these conditions the segments of the both isoclines situated between two peripheral nodules are practically merged into each other (Fig. 1.18, c). This leads to the following, at first glance controversial conclusions: (1) formally this situation looks as if it approaches a deterministic 'one cause – one effect' scheme, since now (contrary to what was before) from any single point of the

phase space we can arrive only to one and each time different point of the merged isoclines' segment; (2) meanwhile, any fluctuations of the A value, however small, will

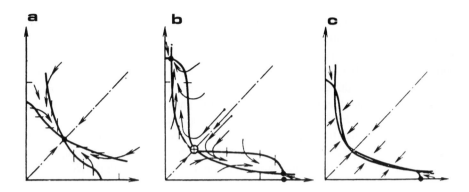

Figure 1.18, a - c. Phase portraits for equation (4) at the different values of A parameter.
a: $A<2$, b: $A >2$, c: $A \cong 2$.

make this point drifting along the whole merged segment. As a result, under these conditions the system will exhibit highly variable and unpredictable behaviour. The final conclusions which we may derive from this situation are as follows:

(a) the deterministic 'one cause – one effect' scheme is no more than a particular case within a larger realm of 'many causes – one effect' and the 'approximately same cause – different effects' relations;

(b) contrary to immediate expectations, the degenerate cause – effect relations provide much more precise and predictable results than one to one relations;

(c) a developing organism, while passing from a non-competent (one potency only) towards a competent (two alternative potencies) state in a way similar to that described by equations *(4)* will inevitably pass via a state of increased variability (non-predictiveness) of its behaviour. That perfectly corresponds to the periods of 'mixed' and variable potencies (developmental fates), as described in the previous section. We can see now that the existence of these properties can be naturally interpreted in a self-organisational way. To this we may also add that the model *(4)* is rather robust, retaining its properties at any power (not necessarily the second one) of the dynamical variables.

Lastly, let us make just a few remarks about possible biophysical meaning of the parameter A. Obviously, it is both rate determining and non-specific, in the sense that it affects equally the rate of production of both metabolites X and Y. As suggested by Professor Chernavsky (see Romanovsky *et al.*, 1984) such a parameter may be a measure of the energetical (oxidative?) metabolism. If we accept this, we must assume that the passage from a non-competent to a competent state (or to any following state associated with the increase of the set of potencies) should imply an increase in energy consumption. The evidences presented by Romanovsky *et al.* support somehow this hypothesis which meanwhile requires much more detailed verification. In any case, the investigation of the model and natural systems of the kind described   may be another route towards discovering the properties of energy consuming (non-equilibrial) dissipative structures.

Generation of autooscillations and the associated regimes
Let us consider now a system of differential equations with two dynamical variables, their T_{ch} differing no less than in order in relation to each other. In the other words, we will introduce a slow variable X and a fast variable Y such that $T_{ch}(X) \cong \varepsilon T_{ch}(Y)$, where ε is a so called small parameter, differing negligibly from a zero value. By such a procedure a set of the dynamical variables themselves (apart from the parameters) will create a bilevelled system. The isocline of horizontals ($dy/dt = 0$) will serve as an attractor, that is as a set of points drawing to it the trajectories of the fast variable (to say more exactly, the fast variables will rapidly, that is, at their own dy/dt rate, be brought into a negligibly narrow ε-neighbourhood of the attractor and later on much more slowly, with the dx/dt rate, move along it). All the rest part of a phase space will be practically 'empty' of the variables owing to their rapid approachment to the attractor.
    The equations which we will explore are as follows:

$$\frac{dx}{dt} = Y \tag{5a}$$

$$\varepsilon \frac{dy}{dt} = - (Y^3 + aY + X) \tag{5b}$$

A qualitative analysis similar to that performed with equations *(4)* shows, that the behaviour of the system will be determined, first of all, by the sign of the parameter a. The isocline of the verticals under any a value   will coincide with the horizontal axis ($Y=0$). However, the shape of the attractor (the isocline of the horizontals) drastically depends upon the sign of a. Under positive a it monotonically decreases at X increases,   thus exhibiting a rather trivial behaviour: from any point of the phase space a trajectory will 'fall' down onto the attractor as we move along it up to the stationary point of the isoclines' intersection (coordinate centre) (Fig. 1.19A). Meanwhile, when $a < 0$ the system's behaviour becomes much more interesting. Now the attractor becomes S-shaped, crossing the vertical coordinate axis in three points. Under these conditions the dynamical variables' trajectory will take the form of a closed loop with two slow branches SP and QR (going along the attractor's parts) and two almost vertical ones, exemplifying the fast variables rate (Fig. 1.19B). In such a way we obtain a regime of *non-linear autooscillations*, which is widely spread in the living nature and can be effectively reproduced in the model systems. Such a reproducibility will not look too surprising if we bear in mind that this regime is extremely stable (robust) both dynamically and parametrically. Its dynamical stability is, indeed, absolute: starting from every point of the phase space we come to the same autooscillation contour. Its parametrical (structural) stability is not so universal, but it is also considerable: any variations of the shape and location of the verticals' isocline which do not bring it outside a "loop" created by the attractor will in no way affect the autooscillatory regime. For example, if instead of $Y = 0$ axis the verticals' isocline will be represented by a wavy line UV (Fig. 1.19B), nothing at

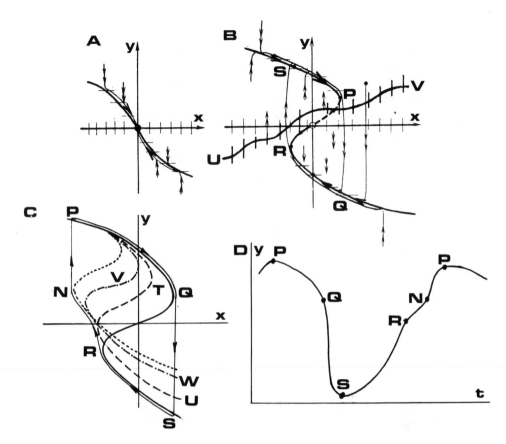

Figure 1.19. Phase portraits showing the emergence and the modification of the autooscillations. A: a non-oscillating regime with one stationary point in the coordinate center ($a > 0$). B, C: $a < 0$. B is a symmetrical autooscillatory limit cycle. C is an asymmetrical limit cycle generated under the assumption that the parameter a is drifting with a rate comparable with that of the dynamic variable. D is this cycle unfolding along a time axis (t).

all will be changed in the system's behaviour. As concerning the values of the parameter a, their changes, so far as they remain within the negative range, will either enlarge or shorten the 'slow' branches, without affecting qualitatively the oscillations' regime.

Let us now ask ourselves what will happen if we allow the value of a parameter a (together with other, as yet non-specified parameters) to drift with a rate comparable to that of the dynamic variables. Suggest (Fig. 1.19C) that we start from the horizontals' isocline to have the shape $PQRS$. Suppose also, that the dynamic variable passes, under the constant parameters' values the way PQS. Then let the parameters' values change in such a manner that the horizontal's isocline will take successively the positions PTU, PVW, etc.. If these changes go at the rates comparable with that of the dynamical variable movement, the latter will pass smoothly from one piece of the changing isocline to another, taking the route *SRNP*. As a result we obtain an asymmetrical, shoe like contour of an autooscillation; its unfolding along the time axis is shown in Fig. 1.19 D. As we shall see later on, a number of biological autooscillations exhibit just this pattern indicating that the parameters are drifted indeed with the dynamic variables' rates. That may point to the reverse dependence of the parameters upon the dynamic variables. Such a bidirectional connection of the parameters and the dynamic variables may be of a great biological importance, not only in the cases of autooscillations.

On the other hand, by varying the position of the verticals' isocline in a wider range than before, we shall obtain two other important modifications of the autooscillations regime. First of all, a regime of the solitary (or waiting) oscillations will be born if the intersection point of both isoclines takes a position outside the attractor's loop. For doing this, one should exchange the equation *(5a)*, by something like

$$\frac{dx}{dt} = f(X, Y) - B \tag{5c}$$

Now the verticals' isocline will become shifted in B units up from the $Y = 0$ axis (Fig. 1.20A). As a result, we shall obtain a stationary (and stable in a restricted neighbourhood) point S, so that for producing each subsequent oscillation the variable should be affected by an X-perturbation, large enough to bring it towards the 'edge' of the loop (as shown by a horizontal arrow coming from S). This regime, similarly to the previous one, can also be modified by the parameter's drift. It easy to see, for example, that the increase of B value during the dynamical variable approaching point S from below will prolong the pulsation, while by diminishing the same parameter's value in the next time period we shall facilitate the reaching of the instability barrier, and hence the production of a new oscillation.

Instead of the 'waiting regime', a so called 'trigger regime' will be generated if the verticals' isocline crosses the attractor's loop in three points (Fig. 1.20 B). Now the extreme intersection points L and N will become the nodules which are stable in a

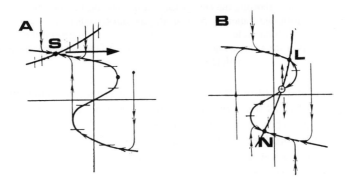

Figure 1.20. Some regimes associated with the autooscillations. A: solitary oscillation (waiting regime) with a stationary point S. B: a trigger regime with two stationary points L and N.

limited neighbourhood, so that after large enough horizontal perturbations a variable can 'jump' from one nodule to another. The trigger regime may be regarded as another, and probably most general model of embryonic differentiation. The main interest of such a construction is that it points to a close linkage between the autooscillating regimes and the passage towards stable differentiation.

From a more general viewpoint the importance of the described models is in tracing the realistic ways of creating completely new (non-preexisting within a given system) spatial and temporal scales. We can now see, that for performing this at least a three-levelled system is required (in an ascending succession the levels are exemplified by a fast variable Y, by a slow variable X and by the parameters). In addition, the dynamical variables should be linked by '+, – ' feedback. Namely, an increase in the fast variable Y causes the increase in dx/dt, (this is a positive feedback link) while an increase in the slow variable X produces instead dy/dt decrease (this is a negative link). Very formally, Y may thus be regarded as an 'activator', while X as an 'inhibitor'. A newly generated temporal scale is exemplified by the period of autooscillations, which is the only well defined T_{ch} within the whole system. As can be shown by algebraic calculations, it is proportional to $a^{3/2}$. Therefore, depending upon a value, T_{ch} can be made as large as possible. Since, however, in the canonical systems of such a kind a T_{ch} for the parameter a is taken in an order greater (up to infinity), the newly generated T_{ch} (the autooscillation period) should be put somewhere in between the slow variable's T_{ch} and the parameter's T_{ch}-s. That means that a newly generated temporal level occupies an intermediate position between both pre-existing ones. Such an arrangement of the new structures (new levels) inbetween the already existed ones is quite typical for the self-organising phenomenae, both in the ontogenesis and in the evolution. Also, as we shall see quite soon, in addition to a new temporal scale a spatial one can be easily acquired, thus breaking the translational symmetry which was beforehand of an infinitely great order. It is actually a real self-organisation that we get from this class of non-linear equations.

Another appropriate term for expressing what is taking place in these systems is a *synphasiness,* or a *coherency.* A coherency is usually associated with the functioning of lasers, emitting coherent light. Meanwhile, in its broader aspect a coherency is one of the most common properties of a large class of unequilibrial (dissipative) structures to which the active (energy pumped) lasers also belong. A passage from a non-coherent to the coherent regime is realised in such systems as a kind of a phase transition after exceeding a certain energetic threshold. In the system (5) it depends firstly upon the sign and then upon the absolute value of the parameter a. As we could see, with the large enough negative values of a we obtain real macroscopic 'jumps' of a system's state, going with a fast variable's rate. If we take into consideration that these jumps mean abrupt and practically synchronous changes in the large collectives of molecules, cells, or other units, a strong synphasiness in their behaviour becomes obvious. Remarkably, the larger is the number of the particles contained in a system, the easier it would be to involve them in a coherent regime (Prigogine, 1980, p.144). Therefore the coherent regimes well may take place in the macroscopic systems, whose number of particles approaches infinity.

Taking into consideration the robustness of an autooscillating and the related regimes, it would be quite strange if the living systems, being definitely macroscopic, were not to exhibit some kind of a coherency. Unfortunately, up to now only a few investigators have associated their interests with this exciting problem. They have already got, nevertheless, substantial evidences indicating the coherency of an ultraweak light emission from the organisms (Popp *et al* eds., 1992). In its broader aspect the problem of coherency in living systems was discussed by Ho (1993). One of main advantages of coherent regimes would be their capacity to produce, under some conditions, immensely fast responses, generalised throughout an entire macroscopic system. Such responses will be impressive most of all in the systems which look as stationary ones and are actually located at the stable nodules of a waiting or a trigger regime: a generalised fast coherent reaction induced in these systems by an adequate dynamical perturbation will look from a deterministic point of view as something quite unexpected.

From determinism to constrained stochasticity: strange attractors, discreteness, and fractals
Here we will touch quite superficially, and only so far as it is related to the developmental topics, a very important branch of modern mathematics dealing with the appearance of randomness and even of chaos out of initially 'deterministic' events. By 'deterministic' we mean here anything precise and predictable, rather than that obeying the 'one cause–one effect' rule. What we have already learned is that such events can be generated in a much more reliable way out of a 'noisy', fluctuating environment with the use of a parametric (i.e. 'holistic') regulation, rather than according to a classical scheme of uniform determinism. We would like now to review quite briefly some of the most beautiful results of recent mathematical investigations, demonstrating that under rather general and easily reproducible (and hence robust) conditions the ordered events that emerge on a background of random fluctuations, can be again transformed into a less ordered state, obeying at the same time certain holistic constraints (a 'regulated disorder'). To show this, let us consider a so called logistic equation

$$\frac{dx}{dt} = aX(1-X) \qquad\qquad (6)$$

Its graph looks in dx/dt, X coordinates like a bell-shaped curve (Fig. 1.21 A-D). This function describes a restricted growth (a growth with saturation), which is almost exponential for small X and approaches a zero rate in the neighbourhood of 1. A lot of different biological processes obey this rule. And our subsequent conclusions will not even depend upon whether we are dealing with the logistic equation proper or with another function, if only it is represented by a bell-shaped curve.

In our further account we will use the function *(6)* for obtaining a series of *reflections* of X upon dx/dt. That means the following. Let us suggest that we take a certain X value and calculate its function, that is, a definite value of dx/dt. Then let us suppose that the latter becomes a new independent variable X which determines the next dx/dt value, etc.. In other words we construct a chain, each of its successive links being determined by the preceding one. Even under as vague assumptions as possible, such a chain possesses some interesting properties, applicable to biology. We can see that rather independently of a kind of a function employed (even a linear function is appropriate), the chain of reflections splits a previously continuous movement of a variable into discrete (and generally unequal) steps of a definite length. Let us consider now a biological (morphogenetic) process which can be approximated by a similar chain: for example, a series of the self-inductions. It is biologically plausible to assume that before responding to a given self-inductive stimulus an embryonic tissue remains 'blind' (refractory) to any next stimulus. Under these conditions the reaction will consist of some number of finite successive steps: a process will be objectively quantified, and a system will behave itself as if getting a finite memory. Hence, a quantification and the memory properties seem to be deeply embedded in the very nature of self-determining systems.

By introducing only slightly more specific constraints we come to even more important results. Let us return to a bell shaped function *(6)* and explore its behaviour under different a values. It turns out that for $a<1$ the chain of reflections, irrespective of its starting point, always brings the variable towards zero value, which is the only stable point (Fig. 1.21A). Meanwhile, at $a=1$ a first (but not the last one!) bifurcation appears, zero point losing its stability and a certain non-zero point r becoming now a new stable stationary nodule (Fig. 1.21B). When a increases further a principally new event occurs: instead of being fixed at a single nodule, the function will fluctuate in a stable manner between two fixed values p and q (Fig. 1.21C). That means that a stable cycle with a

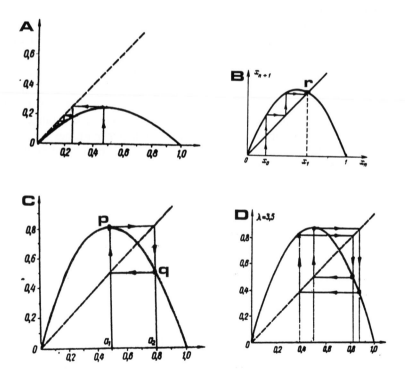

Figure 1.21. Stationary points and periodic events produced by a series of reflections based upon equation (6) under the λ parameter's values increased in A-D order. A: a stationary point is at x = 0. B: a stationary point is at x_1. C: a stationary point disappears and a stable cycle of a 1-st order period is generated. D: no stationary point, a stable cycle obtains the 2-nd order period.

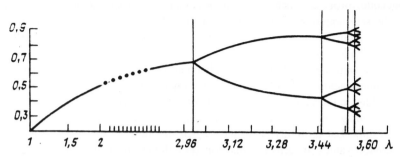

Figure 1.22. A 'tree' of successive bifurcations (doublings of the oscillations periods) as a result of the gradual increase of the λ parameter's value.

period 2 (S^2) is generated. If we continue to increase the value of a, we get successively S^4 (Fig. 1.21D), S^8, S^{16}, etc., cycles, each previous one losing its stability with the appearance of the next. Finally, regular periodicity will be lost completely: what we shall obtain is a bundle of periodic trajectories, each successive one differing slightly from the preceding one (Fig. 1.22).

This example shows us how a mere increase of a single parameter's value shifts a system's behaviour towards stochasticity, and even to chaos. More refined systems of such a kind belong to the category of so-called 'strange attractors': those are the bundles of quasiperiodic trajectories, onto which the locations of any discrete points are ever less predictable, as time goes on; on the other hand, an entire bundle retains its location within a definite region of a phase space. Such a combination of a holistic determination with the unpredictability of the elementary components' behaviour makes the 'strange attractors' and related models in many respects adequate for the biological purposes.

As noticed by Goodwin (1994), biological systems prefer to be located on the verge of a chaotic state, rather than within the depths of a deterministic domain. In the cited book ant populations have been taken as an example, but same seems to be true for the developmental processes as well, including gastrulation (Cherdantzev and Skobeyeva, 1994). According to Kaneko and Yomo (1997) model, if introducing a regime of so called 'open chaos' (that implying the increase of the number of independent variables within a system) the identical oscillating elements acquire an inherent tendency to differentiate, firstly by oscillations' periods and then by amplitudes. This may be a basic mechanism for a non-deterministic (independent from specific inductors) cell differentiation. Inspite of their potential danger, the chaotic states may be really productive!

Chaotic and quasi-chaotic structures briefly reviewed above also have a fundamental property of *self-similarity* at different scales, or, in the other words, *fractal properties*. That means that they look much the same under the different space and/or time resolution. The fractals theory exemplifies another great breakthrough in apprehending the most universal properties of our world. As related to the developmental topics, the main lesson of fractal theory is that the discreteness and scale-invariance of some fundamental properties is quite deeply embedded in the very time–space structure of the Nature, including certainly a living one (Aon and Cortassa, 1996). Correspondingly, a multilevelled structure of the living systems can be better regarded as a mere reinforcement of the overall natural tendencies, rather than a completely new, specifically vital property.

Spatial unfolding of the autooscillating regimes

Such an unfolding (the transformation of auto*oscillations* to the auto*waves*) is achieved, in terms of mathematical models, by introducing a so-called diffusion component. It should not be definitely associated with a real chemical diffusion. Any process, continuously propagating within a space, may be used for this purpose if it fits the following mathematical condition: the diffusion mediated increase of the dynamical variable's value must be linearly proportional to the *second* (rather than first) derivative of this value per space unit. That means that the spatial gradient of the dynamical variable value should be non-linear.

The simplest one-dimensional example of a diffusion mediated wave propagation (a propagation of a 'wave overfall') is described by the equation

$$F(x) + D_x \qquad (7)$$

where $F(x)$ is a non-linear function describing a 'point-located' (rather than space-unfolded) dynamics of the variable x, D_x is a diffusion coefficient and the remaining part of the second term is the second derivative of x per length unit, l. As concerning the point dynamics, a necessary condition for creating a moving 'wave overfall' is to have two asymmetrical stable solutions, as took place in equation *(3)*. Then the moving overfall will attract the successive portions of the active medium towards a more stable solution. This structure, which looks at a first glance rather monotonous (Fig. 1.23) and non-associated

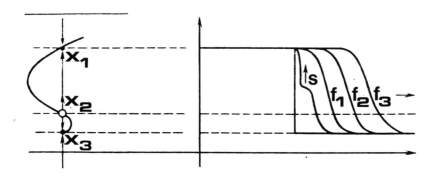

Figure 1.23. Formation and movement of the 'waterfall'. To the left is the graph of a point function $x = F(x)$ and to the right is its spatial unfolding. x_1 and x_3 are stable nodules while x_2 is an unstable nodule. f_1–f_3 are the successive phases of the waterfall's movement. The vertical arrow s indicates the corresponding concentrational shifts.

with a new dimensional scale, is nevertheless a perfect starting point for very important morphogenetic constructions to be later on discussed.

The classical autowaves like those appearing in the Belousov-Zhabotisky reaction can be modelled if combining a diffusion with an oscillatorial, that is, 2-dimensional, point dynamics. The rate of diffusion should be comparable with T_{ch}^{-1} of the oscillations. Under these conditions a new dimensional scale immediately appears.

Stationary dissipative structures
Their history starts from a famous theoretical study of the British mathematician Alan Turing (1952). He invented some imaginary chemokinetic conditions for generating stationary waves from an initially homogeneous medium, affected by the chaotic fluctuations (by a 'white noise') of the dynamic variables (only much later a real physico-chemical model of the similar events had been discovered, see Winfree, 1991). The stationary waves produced may be either quite regular (Fig. 1.24, left frames) or, on the verge of instability, more or less stochastic (same Figure, right frames). They are indeed dependent upon the initial and boundary conditions, but such a dependence is quite far from being deterministic: whereas their very appearance out of a homogeneous state is induced by certain shifts in the dynamic variables, the very *shape and number* of the waves is determined exclusively by the homogeneously distributed parameters values. In the

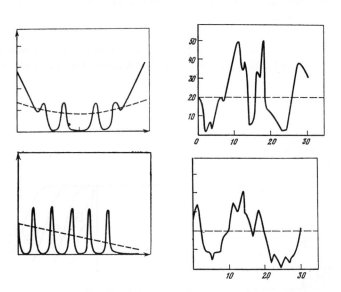

Figure 1.24. The Turing structures (dissipative stationary concentrational waves). In all the frames the horizontal axis is the length of the reactor whereas vertical axis denotes concentrations. Left frames: two examples of the regular Turing structures obtained far from the instability barrier for different slopes of the parameters values (dotted lines). Right frames: examples of somehow stochastic structures obtained in the vicinity of the instability barrier (from Romanovsky *et al.*, 1984, with the author's permission).

other words, the Turing figures demonstrate once again the dualistic subordination of the self-organized systems to the dynamical and parametric regulation and, also, the possibility of the 'spontaneous' (non-Curie) dissymmetrization. Without going into mathematical details, we would like to follow some conclusions from the works of Turing and his successors. Those of a main biological interest are related to the dependence of structures' formation upon the linear dimensions of a 'reactor', in which the corresponding processes are going on. It was found, in particular, that the ability of a system to build up the periodic dissipative structures (to create heterogeneity out of an initial homogeneity) depends in a threshold manner upon the reactor's dimensions: in too small reactors the initial homogeneity remains undisturbed. Above a certain dimensional threshold, a structure of a half wave-length is immediately formed. By continuing to increase gradually the reactor's length, the series consisting of, say, 1, 2, 4, 8,.. half wave lengths, or, under other conditions, of 1, 3, 6, 12,.. half wave lengths structures are generated, while the fractions of half wave-lengths never appear. If, after formation of a certain series of structures, we start to diminish a reactor's length, the thresholds, corresponding to a definite number of structures, become shifted towards smaller dimensions (Fig. 1.25A). In the other words, some kind of hysteresis, or system's memory is taking place. The next important property concerns the relations between so called 'senior' and 'junior' modes. The senior modes are those generated by completely random perturbations, while the junior modes require perturbations having a more definite location. The wave length of the first ones is twice as large as that of the latter. Therefore, while a smaller senior mode's structure contains one half wave and hence contains within no plane of a reflection symmetry (Fig. 1.25 B, *mn*), a corresponding junior mode's structure as containing two half wave lengths possesses such a plane (same frame, *pqr*). That means that the spontaneously formed dissipative structures have an inherent tendency to be dissymmetrical, whereas for increasing the symmetry order more specific perturbations' arrangement is required – a conclusion, which is also against the Curie principle.

Models of structurization implying the differences in T_{ch} or the temporal delays.
The ways for creating structures out of a structureless state formulated by Turing were too narrow: as mentioned above, even in the artificial physico-chemical systems they could be reproduced only under quite specifical conditions. Besides, some properties of the Turing structures greatly differ from those observed in developing organisms. For example, Turings structures do not exhibit, without additional assumptions, the fundamental biological property of a scale invariance (proportionality of structures at different linear dimensions) which forms the very essence of the above described Driesch' embryonic regulations. A simultaneous formation of an entire series of structures obtained in Turing models is also atypical for biological morphogenesis. The question arises of whether one could suggest some more general and biologically reproducible conditions for producing structures, if not out of a completely structureless state (which is also, from a biological point of view, too strong an assumption), then, at least, from a less structured one. It was soon realised that the set of such possibilities was greater than initially expected. For the sake of simplicity, we may firstly divide them into those formulated in chemokinetical and in mechanical terms (although in reality there should not be such a strict subdivision, and systems of a mixed nature are mostly plausible). Addressing now

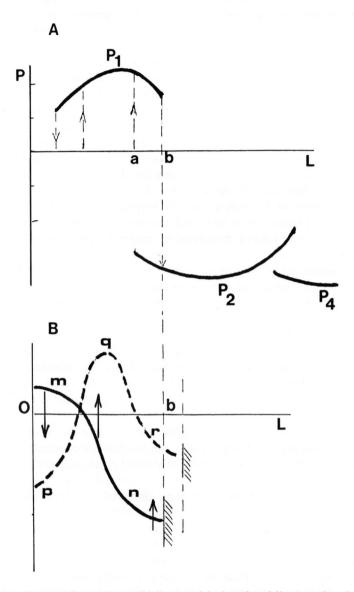

Figure 1.25. Hysteresis properties, senior (solid lines) and junior (dotted lines) modes of the Turing structures. In A the vertical axis gives the amplitudes of the stationary waves measured in half wavelength units. Hysteresis is indicated by the overlapping of the structures (solid arches) generated during a gradual increase (P_1) or a decrease (P_2) of a reactor length. The P indexes correspond to the half wavelengths numbers. B: for a fixed reactor length a senior mode mn can be generated by stochastic perturbations, while the junior mode pqr , its wave length 2 times less than that of the senior mode, requires strictly localized perturbations. Some of them are indicated by the dotted vertical arrows (from Romanovsky et al., 1984, with the author's permission).

the chemokinetical models, we would like to stress, most of all, that they are very much 'playing' with the temporal components.

Gierer and Meinhardt's family of models. This large and fairly popular family of models (Meinhardt, 1982) implied the existence of two different substances, an activator (A) and an inhibitor (I) which diffuse along a 'reactor' with different rates, the coefficient of diffusion of the inhibitor being much greater than that of the diffusion of the activator. As for the 'point dynamics', it is assumed, that the activator catalyses in a non-linear way both its own and the inhibitor production, while the inhibitor suppresses the activator's synthesis. Consequently, similarly to the autooscillation model, a combination of positive and negative feedback is used here. As a result, the sharp peaks of the activator's concentration, alternating with smoother and more extended inhibitor 'clouds' are produced, even out of a homogeneous fluctuating state. In such a way the structural periodicity and other related events can be imitated in their general outlines (Fig. 1.26).

From the biological viewpoint, the importance and the plausibility of the model discussed primarily depends upon whether the main morphological structures of the developing organisms are necessarily preceded, in a one to one manner, by the chemical (concentrational) 'blueprints' (prepatterns). Is it really so? Or would it be more cautiously and realistically to suggest that the concentrational prepatterns, even if taking place at all, are as a rule much more smooth and vague (have much greater symmetry order) than the subsequently arised morphological structures? Such is one of the mostly debatable questions of the modern developmental biology, which shall be discussed later on in this

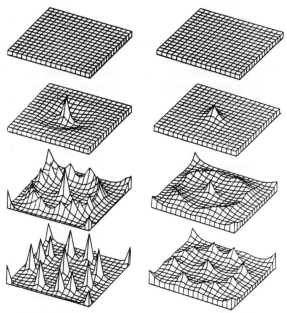

Figure 1.26. Development of the 2-dimensional concentrational Gierer–Meinhardt structures, gradually emerging from a homogeneous state (upper row). Left column: patterns of the activator concentration; right column: pattern of the inhibitor's concentration (from Meinhardt, 1982, with the publisher's permission).

book in greater details. On the other hand, it is possible to regard the Gierer-Meinhardt model as that reflecting some fundamental properties of morphogenesis, not necessarily linked with the postulated chemical prepatterns. Such properties may be associated with: (a) the combination of a short range activation with a long range inhibition and (b) with the differences in T_{ch} -s of the dynamical variables or of the parameters. To the first of these assumptions we will return in the Chapter 2, suggesting that it can be interpreted in quite a different way. And now we shall discuss the second one.

Polezhayev and Ptitsyn (1991) model. The authors consider the formation of periodic circular structures in the bacterial colonies. In their view, the creation of periodic structures of such a kind can be interpreted as implying the action of no more than one chemical variable, a 'mediator', which is produced by the bacterial cells, splits with a constant rate, and induces the transformation of the cells from one alternative state (the active, or vegetal one) to another, anabiotic state. The main condition for increasing the local concentration of anabiotic cells appeared to be to introduce a *temporal delay* in their reaction to an increase of the mediator concentration. This simple assumption looks as biologically plausible and heuristically fruitful: the delays (implying some kinds of threshold functions) are, in any case, widespread, if not universal features of all the biological responses and seem to be largely used by organisms in regulatorial purposes. We shall address this point many times in Chapter 2.

1.4.3. DISSIPATION OF ENERGY AND RELIEF OF MECHANICAL STRESSES IN BIOLOGICAL STRUCTURING

That living systems are far from thermodynamical equilibrium, is out of the question. This is directly evidenced, for example by their continuous emission of a small number of photons throughout the whole spectral range, from ultraviolet to infrared (Popp *et al.*, 1992). Meanwhile, if being under equilibrium conditions, in order to emit light in the red range a body should be heated up to approximately $350°$ K and for emitting in the ultraviolet range up to $1000°$ K! Therefore, in some respects, at least, a living body behaves as a highly unequilibrial one, so that the presence of some kinds of really dissipative structures within is unavoidable. That does not mean, however, that those arising during morphogenesis directly belong just to this category: with some few (although perhaps remarkable) exceptions, even morphological structures just emerged do not disappear immediately after blocking the inflow of energy, as real dissipative structures should do. A much more universal fact (from which we know no exceptions, and which will be discussed in greater detail in Chapter 2) is that all of the morphogenetically active tissues are mechanically stressed, and that *a sudden relief of the stress leads to immediate deformation of the piece of tissue in the same direction as during its normal subsequent morphogenesis.* From this we may conclude that the visible morphological structures which arise during development should be better attributed to the stress-relief, rather than to 'purely' dissipative ones. In more precise terms that means that *the morphological structuring may be regarded as a part of a downhill movement towards a (never reached!) minimum of mechanical energy, this energy being stored beforehand within a given structure.* Such will be one of the central ideas of this book, to be later detailed and substantiated. We shall start on this path by considering some simplified and at the first glance mechanistic, but in fact biologically oriented, examples.

Take an elastic membrane and stretch it in the direction which we shall define as a longitudinal. It will immediately contract itself in the perpendicular (transverse) direction. This is a well-known in mechanics 2-dimensional Poisson's deformation (longitudinal extension associated with transversal compression). If, however, the deformed material is thin and elastic enough, the deformation, instead of going on in a smooth manner, will lead to the formation of a series of longitudinal grooves. Hence, in addition to the dissymmetrization directly caused by the external stretching force, we obtain an additional 'spontaneous' translational dissymmetrization and the creation of a new macroscopic scale (this is exemplified by the transverse diameter of a groove and the distance between the grooves).

Let us explore more closely how this can take place in bodies which may be regarded as mechanically similar to biological tissues. Take a quasi-elastic grid shown by wavy lines in Fig. 1.27, 1 and pull upwards in AA_1 direction the midpoint of its upper surface BAC. As a result the fibre BA will be stretched to BA_1 and the fibre CA to CA_1. The only way for the both fibres to relax towards their initial lengths will be for the point B to move itself towards B_1 (assuming that $A_1B_1 = AB$) and for the point C to move similarly ($A_1C_1 = AC$). This illustrates in the most general way the coupling between longitudinal

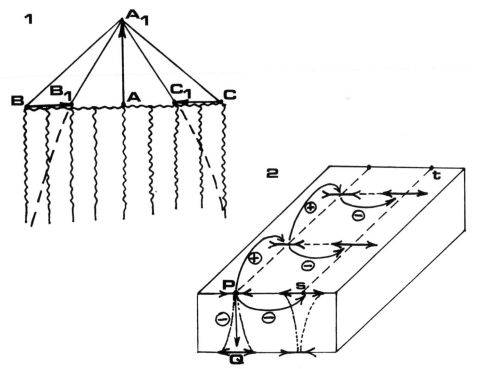

Figure 1.27   Mechanical events associated with a stretching of biomorphic tissues. $\oplus$ indicates a facilitation while O the hampering of the same sign deformations. For detailed comments see text.

extension and transverse compression within biomorphic tissues. One can easily see, that it is based upon the *relaxation* shifts of the parts towards a *local* energy minimum, implying the presence of the pulling force. Obviously, in the absence of this force the accessible energy minimum will be deeper, corresponding now to the undeformed grid.

Now take a large enough piece of a similar 3-dimensional tissue (Fig. 1.27, 2) and suggest that its upper surface is smoothly stretched longitudinally and compressed in the transverse direction. Let us also suggest that owing to some inevitable small inhomogeneities the transverse compression in the vicinity of a certain point P will be greater than in the surrounding regions. It is easy to see that this will facilitate a similarly directed compression in other sections of the same longitudinal axis (shown by converged arrows) and, at the same time, will hamper the compression to the lateral (say, along st line), initiating here, on the contrary, a deformation of the opposite sign, that is transversal stretching. Therefore, the compression zones will be arranged as a series of discrete longitudinal lines situated not too close to each other. Also, on taking into consideration the incompressibility of a deformed material the transverse contraction in P vicinity will cause the extension at the opposite side Q of the piece, and vice versa. Such an increased focusing (resulting from internal mechanical constraints of the most general nature) of a smooth initial deformation will reduce very much the symmetry order of a given system. This is perfectly illustrated by a folding pattern on the surface of an inflated cellophane bag (Fig. 1.28 A). As we shall see later on, these up to now passive focusing tendencies should be largely modulated and reinforced by the active mechanics of the developing embryos.

As a next example of a self-complicating mechanics consider an inflated elastic shell having the shape of a flattened rotational ellipsoid with long semi-axis a and short semi-axis b. It turns out (Martynov, 1982) that if and only if its long semi-axis exceeds the short one by no less than a factor of $\sqrt{2}$, the equatorial belt of the ellipsoid will undergo stretching along the shell's meridians ($\varphi \, ||$) and compression along the equator ($\varphi \bot$). The corresponding stresses will be:

$$\varphi^{||} = \frac{pa}{2S} \; ; \quad \varphi^{\bot} = \frac{pa}{2S}\left(\frac{a^2}{2b^2} - 1\right) \qquad (\, a > b\sqrt{2}\,) \qquad\qquad (8)$$

where S is the thickness of the shell. From (8) we obtain that if $a \leq b \sqrt{2}$ there will be no equatorial compression, however great the pressure force may be. Hence, the stress pattern is largely geometrically dependent. The main interest of this example is that if the shell is not too thick and rigid it will become segregated into a series of equatorially located meridian folds. On deflating the ellipsoid the meridian (vertical) folds will be exchanged for horizontal ones (Fig. 1.28 B) The number N of the folds depends only upon the ratio of a long semiaxis a to the shell's thickness S, being thus the function of a mere geometry:

$$N \cong 4\sqrt{\frac{a}{S}}$$

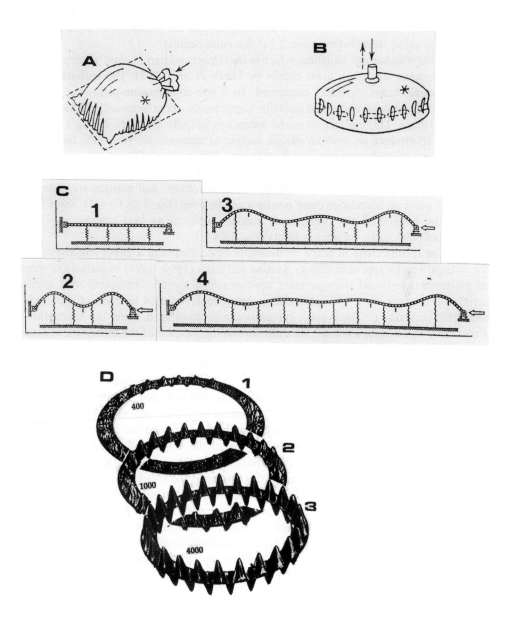

Figure 1.28. Mechanical ways for generating periodic structures. A: a folding of an inflated cellophane bag. B: inflation (solid arrow) of a flattened balloon generates a series of vertical folds, while its deflation (dotted arrow) produces a series of horizontal folds (from Martynov, 1982, with the author's permission). C, 1–4: formation of a series of undulations in a laterally compressed bar connected by elastic springs with a solid undeformable foundation. D, 1-3: spontaneous origin of a regular whorl pattern in an annulus while starting from a random initial pattern of undulations. The number of model's iterations is shown for each of the successive images. This is closely related to the phyllotaxis in flower plants (from Green *et al.* 1996, with the author's permission).

By Martynov's suggestion, this model perfectly reflects the formation of an apical whorl in the green algae, *Acetabularia* (see 2.1.7 for more details).

A similar mechanical situation, which is also closely related to plant's morphogenesis has been recently treated in great details by Green *et al.* (1996). The authors consider firstly a flexible beam elastically connected to a non-compressible plate (Fig. 1.28 C). Even intuitively, it is obvious that in-plane compression of the beam will produce a series of its undulations. This mechanical model appears to be quite similar to Turing's system as being able to produce *de novo* an integer number of macroscopic structures (in this case bucklings) out of random mechanical fluctuations. The number of the structures produced is proportional to the beam's length and also depends upon the border conditions: with hinged beam boundaries the odd numbers of the structures' half periods are generated, while with clamped boundaries these numbers will be even (Fig. 1.28 C, 1-4). Same model can be also applied to the pressurized annulae, reproducing in a realistic way some complicated phyllotactic patterns in plants (Fig. 1.28 D).

Another related trend in shape modelling is associated with a direct employement of the minimal energy approach. Thus, Svetina and Zeks (1990, 1991) modelled the shapes of the membrane-bound homogeneous vesicles which should correspond to a minimal membrane bending energy W_b , the energy being regarded as a function of the average membrane curvature C. The average curvature can be considered, under the assumption of a constancy of a vesicle's volume, as a measure of the increase of the membrane area and, correspondingly, as a result of an increase in the tangential (planar) pressure within the membrane. Some of the model's conclusions are shown in Fig. 1.29. What is, in our view, mostly interesting among them is that an absolute energy minimum, corresponding to a shape 2 (having a symmetry order $2 \cdot m$) can be reached only at a certain, not very high average curvature; for the highest curvature only the local minima are attainable, the corresponding minimal energies being gradually increased as the curvature rises. Meanwhile, with the given C values the energies for the shapes with a reduced symmetry order ($1 \cdot m$) appear to be smaller than those for the shapes of an initial order ($2 \cdot m$). Therefore, for high enough C (i.e., tangential pressure) the trend towards a local energy minimum will be associated with the reduction of a symmetry order, that is, with a complication of the initial structure. The shapes arising are actually biomorphic: they start from ovoidal ones and then approach to those resembling unequal cell division, as observed, for example, during the polar bodies extrusion from an egg cell.

In summarizing, we can see that stress-relieving structures share the main properties of 'purely' dissipative ones. Actually, both of them:

(a) require fluctuations (inhomogeneities);

(b) employ both short- and long-range positive and negative feedback;

(c) can be regulated both parametrically and dynamically;

(d) have some memory about their past;

(e) as a result of (a − c) they are capable of symmetry breakings and the creation of new macroscopic scales.

In a close connection with (a − e), the structures formed in the way above described do not require for their proper mutual arrangement any positional information in *sensu stricto*: that is, no one of the structural elements participating in morphogenesis should 'know' its exact position within a whole (see, in this relation, also Green, 1996, 1996a). It is the undivided whole itself which 'puts' them in their proper places! In a paradoxical way the mechanics brings us back (or, may be, forward?) towards the genuine Driesch'

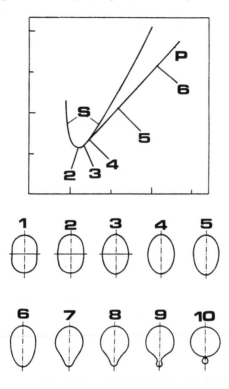

Figure 1.29. Minimal energy figures. Upper frame: a plot of the relative membrane bending energy (vertical axis) as a function of the relative average membrane curvature (horizontal axis). The curve denoted S is for shapes with $2 \cdot m$ symmetry and the curve denoted P for shapes with $1 \cdot m$ symmetry. Numbers correspond to the shapes shown below. Lower frame: examples of the numerically obtained axisymmetrical shapes with minimum membrane bending energy for some values of a relative average membrane curvature. After exceeding a certain average curvature threshold there are $1 \cdot m$ symmetry structures, rather than $2 \cdot m$ symmetry ones, which exhibit local energy minimae (from Svetina and Zeks, 1991, with the authors' permission).

views upon indivisible whole which he himself considered to be compatible only with a vitalism. In any case, practically everything from the general theory of dissipative structures and the related events is true for stress-relieving structures, but not vice versa. The main distinctions of the latter are as follows:
– they are not so much dependent upon the values of the spatially homogeneous variables (for example, chemicals);
– on the other hand, they are greatly dependent upon both local and global geometry (although the dynamics of the dissipative concentration waves also depends upon their curvature) and not so much (if at all) dependent upon the absolute dimensions of a system.

The latter property, directly related to embryonic regulations, will be discussed in more detail in Chapter 2;
– owing to a rapid (of the order of the speed of sound) spread of the mechanical stresses long-range feedback is rapidly and easily established in these structures.

All of these properties make the stress-relieving structures rather suitable for a biological morphogenesis. It will be the task of the next chapter to demonstrate, how a living nature makes a use of them. And now we have to take into mind some new and unusual aspects of the discussed problems.

1.5. A brief excursion into semiotics and systems theory: notions of meaning, context and levels, developmentally oriented.

1.5.1. GENERAL DISCUSSION

In the previous section we directed the reader's attention to a fascinating correlation between a family of quite abstract mathematical notions, created by a few mathematicians, and a set of later and independently obtained empirical data. Whatever may be the basis of such a perfect coherence between the 'pure reasoning' of a human mind and the surrounding Nature, this should stimulate us to extend our exploration of the abstract mental categories in the hope of applying them to developmental topics. We would like now to touch the fertile, but vacillating, ground of *semiotics*, the science about signs and their relations with signified objects. Meaning, value, sign, symbol, metaphor, context – these are some of the key-words of this branch of knowledge, joining together linguistics, philosophy, sociology, mathematics and whatever else you want. We shall start from a notion of a *meaning*.

Why does a strange contour on white paper which we call a letter, or a sound, or an abstract image, sometimes provokes a cascade of human actions, while other somewhat similar figures do not have any such consequences at all? Why is a piece of a paper decorated with a pattern which is itself senseless readily exchanged for food products, clothes, or a piece of land with a house on it, while another similar piece of paper means nothing except of a piece of paper? It is but trivial to argue that the reasons for distinguishing the things which we call signs from other things of the same category are in no way connected with the internal structural differences between the signs and non-signs: nobody in his right mind will try to 'derive' the meaning of an alphabetic letter from its geometry, or the meaning of a sound signal from its frequency spectrum: this is, according to Pattee's (1973) way of putting it, the same as trying to deduce a policeman's authority from his anatomy. We understand quite well that the 'meaning' of a sign is created not by itself, but by a qualitatively different category of events which may be defined as a 'context', 'language' or a 'social agreement'. On changing its context, any sign will either become entirely meaningless or will change its meaning (becoming now used 'metaphorically'). The reverse is certainly not true: a change of a sign's meaning does not as a rule affect at all the context, or the social agreement, with a few exceptions related to quite unstable ('pathological') contexts or agreements (for example, violence and even wars provoked by change in religious or other symbolism).

In any case, the more developed a society and the more perfect a systems of signs will be, the less it will be connected with the primary, ingenious properties of the sign as an autonomous geometrical, physical or anatomical body (compare hieroglyphs retaining some elements of an image with the alphabetical letters entirely lacking such elements; or the warriors of the primitive tribes which should be physically strong with the Pattee's 'anatomy-less' policeman). Moreover, the physical and other effects which may be initiated by certain signs within a certain context are, of course, completely incomparable with the same signs inherent (out of a context) properties.

All of these almost trivial considerations are required for extending and elaborating the notion of levels. A context, a language, or a social agreement can obviously be regarded as an upper level in relation to that of the single signs: the characteristic times and the spatial extension of the former are many orders greater than those of the signs proper: it is among our main social pursuits to extend as far as possible the areas and the time periods of the validity of constitutions or of agreements, their too rapid change being estimated as social catastrophes. By attributing these notions to the concept of levels we should considerably affect and enrich the latter's content.

Up to now, while considering the multilevel systems we have associated the upper level(s) with the parameters, and the lower one(s) with the dynamical variables. Within such a strictly mathematical (or physicalistic) framework the descending interlevel relations are firmly associated with, and reduced to the parametrical regulation. This kind of regulation could much affect the lower levels dynamical variables by vectoring them in a phase space, retarding their rates etc.. However, in the physicalistic multilevel systems the very dynamics, or the energy (a cause of the system's activity) is always resided on the lower levels, compared with those involved in the parametrical regulation. In the semiotic aspect, meanwhile, an upper level units behave in a quite different manner: they do not any longer use the lower level as a source of the dynamics, but instead borrows the latter from some independent source. In semiotic systems, what is transmitted from a lower level to an upper one may be called information, or, much better, single units of information (letters, words, etc.), these units remaining completely dumb until being interpreted at the upper level.

Which model of a level is closer to a biological reality, the physicalistic one closely associated with a parametrical regulation of lower level dynamics, or a semiotic one, implying a context-dependent interpretation of incoming informational units? Most probably the truth lies in between, although organic evolution seems to move generally from physics to semiotics. In any case, with the progress of biology a naive reductionist believing in the 'internal power' of signalling factors (be they non-specific external agents, or specific molecules, including direct products of the activity of genes) gradually gives place to the more realistic view of a largely context-dependent character of their effects (Goodwin, 1985; Nijhout, 1990; Strohman, 1997).

On the other hand, it would be flippant to claim that the signalling molecules are mere signs in the sense that their molecular structure, or, say, atomic properties of Ca^{++} ions, play no role at all in the emergence of their signalling properties. Probably the biological situation can be best of all compared with the hierogliphic stage of a written language.

Some additional problems arise in relation to a context-dependency of the genetic factors. One might well say that a context itself (i.e., the entire enzymatic machinery, micro- and macroenvironment, including morphology, etc.) has been created to a great extent with the help of genes. Yes, operationally it is so, but to assert this is the same as reducing, for example, the state constitutions or social agreements to a set of letters used for writing them: obviously, the meaning of the documents will not be altered even if we change the letters! Returning to the genetic factors, we may consider them as including no less than two different scales entities. The first is exemplified by single genes (DNA loci) switching on and off during development; those are short term and largely context-dependent dynamical variables, their characteristic time not exceeding some several dozens of minutes. The second relates to a species-specific genome taken as a whole. Its characteristic time can be taken as being reversely proportional to the mutation rate. For one of the most rapidly mutating Metazoa species, Drosophila flies from the Hawaian islands, it takes about 20,000 years, or 40,000 generations for accumulating 1% of genetic differences and losing breeding capacities among the lines (Raff and Kaufman, 1983). When viewed in such a light, the genome looks like one of the most stable developmental parametrizators and should be thus attributed to one of the upper (if not the uppermost) levels.

1.5.2. STRUCTURALLY DYNAMICAL LEVELS

Definitions
We will define a structurally-dynamical level as *a set of dynamical structures to which we can ascribe more or less definite linear dimensions (L_{ch}) and characteristic times (T_{ch}), sharply enough delimited from those of other levels.*

    A set of levels arranged according to their L_{ch} -s and T_{ch}-s sequences we will define as a multilevels' hierarchy. Obviously, the idea of such a common hierarchy could be regarded as reasonable only if we could prove that the both its members are really changed in parallel with each other, that is, if the L_{ch} -s and T_{ch}-s shifts are positively correlated. Happily, this was really demonstrated for a large number of quite various dynamic processes, their L_{ch} -s and T_{ch} -s ranging from 10^{-10} to 10^{-1} m and from 10^{-13} to 10^{4} s correspondingly (Aon and Cortassa, 1993). Such a correlation appeared to be linear if plotted in a double logarithmic scale what means that the rates of the both members' changes were proportional to their absolute values. Remarkably, these 'weighted rates' were abruptly changed just at the level of a Frankel's barrier (see p. 45), that is within the micrometers - minutes dimensional scale. All of these properties point to a highly regular and a holistic character of an entire L_{ch} - T_{ch} hierarchy.

    Sometimes the L_{ch} - T_{ch} parallelism looks as being violated in the sense that the entities of a relatively small dimensional scale seem to be associated with relatively large and variable T_{ch}. For example, the individual cells' cycles may extend from several dozens of minutes to several hours, days and even years (the latter may be exemplified by a so called diakinesis which preceds maturation divisions in germ cells). Actually, however, such a violation is false. What should be taken here into consideration is an above mentioned capacity of an upper level to slow down the lower level variables rates, or to "enslave" them, in Haken's (1988) expression. In the case of delayed cell cycles the role of the enslaving factor is played by an extremely slow hormonal dynamics of the entire organism, that is by the events coming from a much higher level. Therefore, for evaluating a real, or 'pure', given level T_{ch} we should take into consideration the most rapid dynamics exerted on this level under the conditions of its maximal isolation from the upper levels. After these remarks we may suggest the following draft of the levels hierarchy, oriented in an ascending direction.

1. As the representatives of a lowest level, we take *the macromolecules, able to absorb and dissipate energy*. Their main function is in slowing down the energy relaxation rate, thus making possible the transformation of their excitation energy into that concentrated on the mechanical freedom degrees (Bluemenfeld, 1983). Their L_{ch} is about 10^{-8} m (this is a diameter of a large protein molecule) whereas T_{ch} (the relaxation time) ranges from 10^{-3} to 10^0 s.

2. The next level to above is represented by *non-covalently binded molecular associations still to be placed below the Frankel's barrier*. Those are, for example, actins and actin-binding proteins, which directly perform mechanical work. Their L_{ch} -s vary from 10^{-7} m (myosine molecule) to 10^{-5} m (acrosomal tube), while their T_{ch} can be estimated as about 10^0 s (this is the polymerization time of an average size F-actin molecule).

3. Then we come to the *intercellular structures situated above the Frankel's barrier* (higher orders cytoskeletal formations, vesicular associated processes etc). The vacuoles diameter is in 10^{-6} - 10^{-5} m range while their recycling time lies in 10^1 - 10^2 s range.

4. Next level corresponds to *entire single cells*. Their diameters are about 10^{-5} m while the T_{ch} -s are estimated as the minimal ones required for contact cell-cell interactions (see 2.3.5), in particular the establishing of new cell-cell contacts and a break of the old ones. This takes about several dozens seconds, that is $\cong 10^2$ s.

5. Even higher level is exemplified by *multicellular collectives*, behaving as common regulatorial and morphogenetic entities (morphogenetic fields). To these we attribute entire early embryos as well as separate rudiments of more advanced embryos which manifest Driesch' regulations. Their diameters are no more than 10^{-4} - 10^{-3} m. We take their T_{ch} -s as the time periods between the first delimitation of a given structural unit from the surrounding tissues and its own splitting to the next order units. This is about 10^2 - 10^3 s.

6. This level we attribute to the *individuums, that is entire organisms together with their life-times*. Here L_{ch} -s are of 10^{-3}-10^1 m range while T_{ch}-s are about 10^5 - 10^8 s.

7. Here we outline *populations sharing common genotypes*. Their L_{ch}-s can be estimated as a   diameter of a total 'condensed' population biomass while the T_{ch}-s must be regarded as a minimal time which is required for two ecologically separated parts of a population to become genetically isolated from each other. As mentioned before (p. 70) , for Hawaian flies this is $\cong 10^{11}$ s.

8. At last, we take the total biosphere. Its L_{ch} should be estimated in the same manner as that of the previous level and hence exceeds the latter enormously, while this level T_{ch} corresponds to the time passed from the moment of, say, Eukaryota origin ($>10^{17}$ s).

(Note that an entire human's history with all of its space unfolded monuments should be placed just between the levels 6 and 7: as usually in the self-organizational systems, a newly arised space/temporal level is situated just inbetween the ones already existing!)

Some general properties of the levels interactions
Inspite of such an enormous diversity of levels an attempt to outline some fundamental and invariable features of their "vertical" interactions may probably be not at all hopeless. Let us consider a pair of interacting levels, an upper and a lower one. What we have learned up to now was that in the physicalistic terms a lower level provides a fast dynamics while an upper one exerts parametric regulation, generally retarding and vectoring this dynamics. We mentioned also, that in semiotic terms a lower level produces the elementary "information units", while the upper one is responsible for their interpretation.

To this we would like now to add a suggestion, that the activities of the coupled lower and upper levels are exerted heterochronously, rather than simultaneously, thus creating repeatedly biphasic interaction loops: each one loop starts from a relatively brief phase of energization, associated with a decrease in the symmetry order of the lower level units; at the same time a symmetry order of the upper level structures can remain the same or even increase. This phase is followed by a more prolonged one associated with the energy discharge (relaxation) and the transmission of a dissymmetry in the ascending direction, towards an upper level; the lower level units are now returned towards a greater symmetry by some more or less ordered (vectored) movements. We assume also, that the main function of the descending parametric regulation is to retard these relaxation processes and to bring order into them.

In semiotic terms, the first phase is associated with the 'production' of a new raw information (sets of signs) at the lower level while the second phase is associated with its interpretation on the upper level.

In a mostly straightforward way such a biphasiveness is revealed by the macromolecular and supramolecular levels' events. In the first of them, a combination of a fast excitation (the rate of vibrational movements of protein molecules has values of the 10^{-9} s^{-1} order) with the long relaxation time (exceeding the excitation time by up to 9 orders) is crucial for the functioning of the 'molecular machines' (Bluemenfeld, 1983) and is probably also required for photons storage (Slawinsky, 1988). In supramolecular mechanochemical devices (actomyosin molecules) the energization (associated with ATP hydrolysis) also takes but a minor fraction of the entire working cycle time, while the mechanical work is produced during a more prolonged relaxation phase. About the entire cells level we know quite little, but nothing contradicts our suggestion. As to the level of morphogenetic fields, one of the main aims of the next chapter will be to show that the loops of a similar structure, including a relatively fast generation of mechanical stresses and their prolonged discharge (which we will call quasi-relaxation), create a very essence of the 'working cycles' of the structures generation. These loops may have also a semiotic interpretation, since a rapid phase of stress generation corresponds to the creation of a 'morphogenetic information' about the next period of development while during a quasi-relaxational period this hidden information is translated into visible shapes. Similarly, a living cycle of an individuum implies a brief period of increase in the dissipative function (heat production, oxygen consumption) followed by its gradual decline over the rest of the life period (Zotin and Zotina, 1993). Same dynamics takes place at the population level (Kortmulder, 1994) and, probably, during the evolutionary formation of new taxa. Addressing the evolution of multicellular animals, one may recall indeed that it started with the Precambrian Explosion, producing the representatives of almost all animal phylae within no more than in about 20×10^6 years, which corresponds to about 2% of the entire evolutionary time; all of the subsequent evolution can be considered as no more than a further detailing within as yet established taxonomic field. Are not the Precambrian events a kind of a real explosion of an evolutionary 'energy', later on gradually relaxing? If there is something reasonable in this suggestion, all of the biological biphasical loops taken together create a single giant fractal, embracing a time range of about 20 decimal orders!

1.6. Summary of the chapter

We have started from comparing the approaches used for interpreting the structuring processes in physics and biology. Contrary to the first impression, they have much in common. Being originated from the same source (ancient Greek philosophy, reaching its highest point in the Aristotelean conceptual system), both of them have been later on largely affected by the reductionist ideology of uniform determinism to come more recently by almost independent ways to quite another set of concepts implying a far going delocalization of the cause–effects relations and the possibility of a self-organization ('spontaneous' formation of the new entities out of a less structured state). We reformulate the main problems and the empirical tasks of developmental biology in terms of symmetry theory, as exemplifying the broadest possible conceptual base. We trace the changes in the symmetry order, either in a physical or a phase space, firstly of the visible (momentary) structures and then of the developmental fates, the latter being considered as multidimensional vectors in the physical and phase spaces. Generally, the symmetry order of both the visible structures and of the developmental fates is decreased as a development proceeds, although the reverse processes are also possible on a minor scale. At the different structural levels the symmetry changes may go simultaneously in opposite directions. Some of the dissimetrization processes in the developing organisms obey the classical Curie principle, while others (Driesch's embryonic regulations and the related events) do not: they imply holistic principles and resemble in some important aspects (but not in all) the properties of inorganic dissipative structures.

In this way we arrive at a concept of morphogenetic fields. In our view these should be considered as a hierarchy of interdependent non-linear fields of the different space-temporal levels. To deal with such fields we should be familiar with the fundamental notions of a self-organisation theory. Among those we discuss most of all the principles of a dynamical and parametric regulation, different kinds of stability/instability and the ways of creating new space-temporal levels. We review the ways for creating the regimes of autooscillations and autowaves (as well as some other related ones) both in deterministic and stochastic versions. Returning to biology, we suggest that the living structures are situated as a rule on the verge of stochasticity. We formulate also the notion of the stress-relieving structures which share some properties of the dissipative structures but are not completely identical to them. We suggest that the morphogenetic structures belong just to this category.

Lastly we describe in physicalistic and semiotic terms the main properties of the different space-temporal levels relevant to developmental biology and give a tentative list of these levels. Within each elementary interaction act of the two different levels we distinguish: (1) a relatively fast phase of energization, reduction of the symmetry order of lower level units, coinciding with the production of a raw information on this level, and (2) a more prolonged phase of energy relaxation, transmission of a dissymmetrization towards the upper level, and the interpretation of information at the latter.

CHAPTER 2
A HIERARCHY OF DYNAMIC STRUCTURES IN DEVELOPING ORGANISMS, AS TRACED IN AN ASCENDING ORDER.

Now we address ourselves to real developmental processes, while continuing to keep in mind the explanatory approach outlined in Part 1 with the hope of using it again. Within such a framework the concept of a dynamic structure will remain central for us. Meanwhile, in this part of the book the dynamic structures associated with developmental events will be treated more or less in isolation from one another, without taking into consideration their natural space-temporal arrangement. In terms of our 'new land' allegory (see Introduction) what we intend to develop in this part of the book will be just a list, or catalogue, of geomorphological elements, rather than a geography and geology proper – those will be left for the next part. And we shall present this catalogue in an ascending order, attempting to show that the dynamic structures do indeed expose a coherent multilevel hierarchy. We start from a supramolecular level and move upwards, via the single cells' level, to that of the large multicellular collectives. For each successive level we will be most interested in its self-organizational capacities, as manifested by symmetry breaks, strong reactions to small perturbations, oscillations etc.. Where possible, we will discuss a degree of thermodynamic non-equilibricity of the processes involved. One of our main goals will be to explore those feedbacks which can provide self-organization both within each level and those acting between the different levels. Necessarily, our review will be rather incomplete and scattered, partly owing to gaps in our present-day knowledge but also because of the impossibility of covering all minor details. For example, a description of supramolecular events will be condensed and will not try to substitute for the numerous text-books in this field. On the other hand, our account can be characterized as very mechanically oriented, that is, concentrating on the mechanical forces (or, better said, stresses) acting in embryonic tissues, as well as deformations (strains), which they produce. In terms of a multilevel approach this means focusing our interests on dynamic events with their L_{ch} and T_{ch} much greater than those of the molecular processes (because classical mechanics deals just with *macroscopic* shifts, deformations, and forces). Such a focusing is associated with our belief (for which empirical support has been increasing) that the developmental *macro*events do really play an indispensable role in providing fundamental morphogenetic feedbacks, rather than being only blind end results of the molecular level events. To the professionals the mechanics which we shall use may look rather uncomplete, since in most cases we will be unable not only to measure stresses with reasonable precision, nor even to separate them adequately from the strains which they produce. Remarkably, those few cases where both the procedures can be accurately performed are not among the most interesting ones from the developmental point of view. On the other hand, some purely qualitative statements, looking at first glance rather imprecise and vague, will be most important for us.

Furthermore, we shall also introduce a distinction between passive and active stresses, one that is practically unknown in classical mechanics.

2.1. Mechanical stresses, strains and the supramolecular devices, that produce them.

The supramolecular level is directly involved in chemomechanical transduction, that is in the transformation of a metabolic energy into a mechanical one. This process is crucial for morphogenesis insofar it is reasonable to regard the latter as the 'active mechanics' of a developing organism. As we hope to demonstrate later on in this chapter, some small part of the mechanical energy produced (which is, in some estimations (Selman, 1958) about 10^{-6} fraction of the total metabolic energy) is stored in embryonic tissues in the form of regularly arranged mechanical stresses, their life times ranging from a few seconds to several dozens of minutes or longer. These stored (temporally extended) stresses are integral parts of the higher levels' (cellular and supracellular) dynamic structures. Thus, before passing directly to supramolecular events, we need to discuss some fundamental properties of the mechanical stresses.

2.1.1. STRESSES AND STRAINS, ANIMATED TO A CERTAIN EXTENT

In mechanics a stress σ is defined as the ratio of a force F to the cross-sectional area S of a body to which it is applied:

$$\sigma = \frac{F}{S} \ , \tag{9}$$

or, more exactly,

$$\bar{\sigma} = \frac{F}{lim S} \qquad (S \to 0).$$

Such a reference to the cross-section of a body means that an externally born force has been effectively internalized, or 'inserted' within a body, thereby becoming able to shift its particles from their initial (quasi-equilibrial) mutual positions and increasing the potential energy of the body. These shifts are reflected in the corresponding strains, that is in the ratios of the resulting size changes to the initial size of a body. A stress/strain proportionality is called a modulus, while the reciprocal relation is called a compliance. These two latter values are the measures of a body's resistance to the stresses (that is, its stiffness, rigidity, etc.) which we shall consider in most cases as a constant one and hence deserving no special attention (although sometimes, an especially as related to the plant tissues this is far from being so). In order for strains to be extended to a macroscopic (in relation to molecular dimensions) scale, mechanical continuity of a body is required, or, in other words, the existence of sufficient mechanical bonds between its particles. Already these basic and at first glance simple definitions, when applied to biology, lead to serious ambiguities. First of all: what is an adequate measure of a cross-section of a biological body, say a cell? Should we take for our estimations a total cross section of a cell, as viewed in the optical microscope or it will be more adequate to consider as mechanically loaded only the intercellular contact zones and the intracellular fibrillar structures anchored to these zones and creating but a small fraction of an optically visible 'cross-section'? The modern estimations of the tension forces self-generated by a single cell vary in two orders range, namely from 10^{-7} N (Ingber et al., 1993) to $\approx 10^{-5}$ N (Jones et al., 1995). By dividing these extreme values to a total average cross section of a cell (which we take as 50×50 mkm = $2.5×10^{-9}$ m^2) we obtain the resulted stress values lying between 10^2 and 10^4 N/m^2. Similar figures are given in Bereiter-Hahn's review (1987, p.10). Among those we can refer to James and Taylor's evaluations of the gastrulation-associated stresses in *Triturus* embryo: those are $3.8×10^4$ N/m^2. This is a rather moderate value, roughly equal to the stretching stress of a 1cm^2 diameter rope loaded by the weight of 100 gramm.

If, meanwhile, we divide the same forces values to the sum of the cross sections of the tension-bearing actin filaments extended throught a cell's body, rather than to a total cross section of a cell, the denominator will look something like $2.5·10^{-17}$ m^2 × 10^3 (the first value gives a cross section of an individual actin filament while the second one the approximate number of the filaments per cell). As a result, we get now enormously great stress values ranging from 10^7 to 10^9 N/m^2 or, what is probably more instructive, a force of about 10^{-5} – 10^{-3} dyn per actin filament. The effects of such forces on the molecular level cannot be neglected since they are comparable with the forces of intermolecular attraction (see Forgacs, 1995; Jones et al., 1995 for further discussion). And now we have to return to the elementary mechanics.

If after the removal of a force (or the destruction of a mechanical bond through which it was applied) the deformation (strain) has disappeared more or less completely and 'fast enough', then the body is called elastic, in the sense being capable of storing within it the potential energy of the deformation without substantial dissipation. A strain discharge is now called the mechanical relaxation. This notion, as applied to developing organisms, will also be the subject of substantial modification, dealing mostly with the characteristic times of relaxation. In particular, we will speak later on about the important category of "quasi-relaxational" processes.

On the other hand, a substantial retention of the deformation after the removal of force indicates that most of the deformation energy has been dissipated beforehand into heat: such a body exhibits plastic (viscous) properties. We will argue that most morphogenetically active tissues exhibit combinations of both elasticity and plasticity, each of these properties being linked with the different space-temporal scales of the deformations and the different categories of the cellular and extracellular structures.

Before formulating the main kinds of stress–strain relations, we have to introduce and explicate the notion of the active versus passive stresses, which is uncommon for classical mechanics, but is, in our belief, of a fundamental importance for morphogenesis.

What we will denote as the *active* stresses will be those produced by a chemomechanical transduction, taking place in a *given area* of an organism (tissue, cell) and at a *given ('present') moment of time*. The *passive* stresses are, on the contrary, those produced either 'outside' of a given point, or 'before' the moment of present time. As a rule, the passive stresses are much more smoothed (delocalized) within both space and time as compared with active stresses.

At a first glance, such a distinction between the active and passive stresses may look like quite an arbitrary one. We hope to show meanwhile that this is not the case. Our main argument is that the distinction can be experimentally tested. For doing this, we made precisely localized dissections of embryonic tissues (Fig. 2.1 A-F) both under normal physiological conditions and under those which inhibited, more or less crudely, the energetic metabolism (from a moderate cooling of a sample up to the use of oxidative inhibitors and even fixatives) (Beloussov et al., 1975). Such a treatment could be performed at a definite Δt time period before the dissection. The post-dissecton deformations reproducible after such an inhibiting treatment were considered as passive ones, Δt measuring a 'temporal depth' of the passivity. That means that in the given cases the stress generating metabolic processes could be at work only prior to the Δt time period, so that the stresses produced have been stored in a given region (most probably, as elastic tensions) for no less than Δt. On the other hand, the post-dissection deformations that can be observed only under physiologically normal conditions should be treated as active, that is requiring the metabolic energy which is produced at the given moment of time (or, better to say, during a not too long 'present-time' period). Consequently, the notions of the passive and active deformations are intimately linked with a discrete space-temporal structure of the studied objects and with what may be called a 'morphomechanical memory' (see below for more details).

An additional related method for distinguishing the active and passive stresses is in estimating the lag periods and the durations of the post-dissection tissue deformations. Those created by the active stresses have a finite lag period and are developed gradually (say, within a few minutes). In contrast, the passive deformations, mostly elastic in their nature, are exhibited at their full scale 'immediately', that is within fractions of a second after a dissection and without a measurable latent period. We were able to detect both passive and active deformations in the embryos of quite different species, including hydroid polypes (Beloussov, 1973), sea urchins (Beloussov and Bogdanovsky, 1980), amphibians (Beloussov et al., 1975, 1994) and chicken embryos (Beloussov and Naumidi, 1977). Some of the passive deformations are shown in Fig. 2.1. Among them, frames A - C illustrate fast contractions of the outer (apical) cell surfaces (just leading to the bendings of the dissected pieces) as well as less visible but always revealed by microscopic examination similar contractions of the transversal cell walls. Similar deformations are shown in frame E, the transversal cell walls contractions being here more pronounced. Meanwhile, frames D and F point to the existence of pressure stresses within cell layers, as indicated by the overlapping of the dissected edges. Later on we will numerously discuss these and related cases.

In one case (bending of a neuroectoderm, as illustrated by Fig. 2.1 A) we were able to estimate roughly the force of the passive deformation. For doing this, we have measured a minimal load which, while being put onto a dissected tissue piece was enough for bringing it back to the initial non-deformed position. Such a load more or less uniformly spread throughout the entire dissected neural plate appeared to be equal to 3×10^{-2} mg $\approx 3 \times 10^{-2}$ dyn. Interestingly, this perfectly coincides with the minimal force value preventing neurulation movements in Triturus embryo: by Selman's (1958) estimations it was 4×10^{-2} dyn.

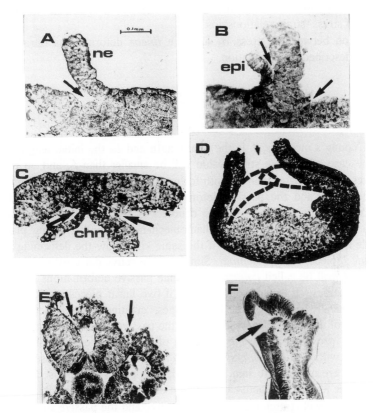

Figure 2.1. 'Immediate' deformations of the dissected tissues which indicate the existence of passive (pre-existing) mechanical stresses of tension or compression in embryonic tissues and their correlations with a subsequent normal development. A-E: frog' *(Rana temporaria)* embryos. F: an early (still non-subdivided) apical rudiment of a hydroid polyp, *Dynamena pumila* (for its detailed description see Chapter 3). . Arrows show the locations and directions of dissections A: a neuroectodermal layer (*ne*) from an early neurula embryo, separated from the underlaid tissues. B: a similar operation, but in addition the outer part of the neuroctodermal layer (so called epiectoderm, *epi*) has been separated from the inner one. C: dissection of a chordomesodermal rudiment (*chm*) from the overlaid neuroectoderm. The extensive and 'immediate' bendings of the dissected parts non-suppressed by metabolic inhibitors have been considered as relaxations of the pre-existing elastic tension stresses. At the same time, they perfectly imitate, in a very much accelerated way, the normal morphogenesis of the same embryonic parts. D: overlapping of the dissected edges of a mid-neurula dorsal roof indicating the existence of pressure stresses (dotted contour shows how the dissected edges should look after their imaginary superposition). E: dissection of a neural tube soon after its closure (left arrow), followed by the dissection of its right wall along a mid-plane (right arrow). An enormous pulling apart of the dissected parts indicates the existence of both the radial and tangential stresses within a neuroepithelium. F: deformation of a dissected distal part of a *Dynamena pumila* apical rudiment. Besides an obvious extension of a dissected part of ectodermal layer (as compared with the length of the naked region), one can see its relaxational lifting up and a folding. Similar folds will appear in the normal samples several hours later, indicating again a 'prognostic' role of the rapid deformations and the complicated nature of the pre-existent stresses. For all the cases the time period between the operation and the fixation of a sample do not exceed 0.5 min.

After these biologically oriented comments we can now return to elementary descriptions of stress–strain relationships.

The relation between a stress σ_E and the *linear* deformations (strains) which it produces are described by an equation

$$\sigma_E = \frac{E(L - L_o)}{L_o} \tag{10}$$

where E is Young's modulus, L is the final length and L_0 the initial length of a sample, which in the case of positive stress value will be smaller than L, and, in the case of a negative stress value greater than L. Both for positive and negative values of σ_E, we will consider the active and the passive deformations separately. The active deformations associated with an increase in L will be denoted as *extensions* (the extending force is assumed to be generated owing to a chemomechanical coupling taking place *within L* piece), while the passive deformation of the same kind (now the deforming force being arisen *outside L* piece) as a *stretching*. Similarly, the active deformations associated with a decrease of L will be called *contractions*, while passive deformations of the same sign are called *shrinkages*. It is quite obvious that if two linear pieces, an active and passive one, are mechanically bound by their lateral edges, creating a closed mechanical contour, the contraction of an active piece will stretch the passive one (producing within it what we call *tensile stresses*), while the extension of an active piece will shrink the passive one, producing within the *compressive*, or *bulk*, stresses: in both cases the active and passive stresses are of opposite sign. If, instead, both pieces are apposed onto each other along their whole lengths, the active contraction of one piece will shrink the adjacent one, while its extension will stretch the latter – the active and the passive stresses will now be of the same sign. Throughout this whole book we will denote passive stresses as single line double headed arrows and the active ones as two lines double-headed arrows. Extension and stretch will be indicated by diverging arrowheads, while the contraction and a shrinkage by converging arrowheads.

Actually, only a few biological structures can be considered, even in a rough approximation, as one-dimensional. Two-dimensional structures (which are denoted in mechanics as shells) form a much larger group. However, for our qualitative approach it will be enough to describe their deformations as combinations of linear ones. The only important addition relates to the coupling between linear deformations taking place in two mutually perpendicular directions: stretching in one of the directions is, as a rule, coupled with the shrinkage (compression) in the perpendicular one. In elastic bodies this is known as Poisson's deformation. In biological samples another (and probably the main) reason for such a coupling is added: it is the incompressibility (volume constancy) of most (although not all) of the tissues.

Voluminous stress-strain relations are described by a relation

$$\sigma_v = \frac{M(V - V_o)}{V} \tag{11}$$

where V_o is an initial and V is a resultant volume, while M is the compression (bulk) modulus. We use the word *inflation* for the stress associated with the an increase of volume, while the word *deflation* is associated with a decrease of volume. Diagrammatically the inflation of a given tissue compartment will be indicated by a filled triangle while deflation is indicated by an empty triangle. No special terminology will be used for distinguish the active and passive volume deformations, although both do occur.

Finally, we need to distinguish *shear* stresses and strains, their relations being expressed by the formula

$$\sigma_G = Gg\alpha, \qquad\qquad (12)$$

where G is a shear modulus and α is the shear angle, being a measure of mutual shifts of two parallel structures. One of the most important mechanochemical devices, the actomyosin system, produces, due to the mutual sliding of the antiparallel actin fibers just shear stresses/strains. Some of the macromorphological stress-strains, to be described below, are also in the nature of shearing. In most cases, however, we shall trace only those consequences of the shear stresses which can be represented as linear ones.

Now we have to discuss an important point which, inspite of its simplicity, may be a source of misunderstanding.

2.1.2. A CONDITION OF A MECHANICAL EQUILIBRIUM

Any 'material point' M of an embryo, be it an intercellular organelle, a cell, or a multicellular collective, will be connected with other material points by a number of mechanical links. If at a certain instant of time it has been shifted from its initial position towards a point N, we say that a force F acting along direction NM, has been applied to M. We know, meanwhile, from mechanics, that any unbalanced force causes acceleration. On the other hand, the observations made on the developing samples teach us, that

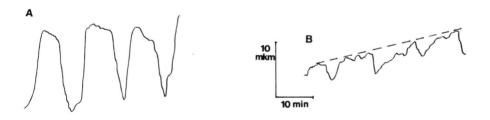

Figure 2.2 A,B. Two records of the periodic extensions–retractions of a growing tip of a hydroid polyp, *Dynamena pumila* (see Part 3, section 3.7. for more details). Note that the records mostly consist of straight lines pieces, indicating that the movements are going on with abruptly changed constant rates. The averaged graph of a linear extension (dotted line in B) is also a straight line (original)

practically all of the morphogenetic movements proceed at fairly constant rates, that is without accelerations: even when the rates are periodically changed, the periods of their changes are negligibly small in relation to the constant periods, so that the plots of the rates of movements look like combinations of straight lines, rather than smooth arches (Fig. 2.2). This can take place, however, only if the vector sum of all the forces applied to a material point is equal to zero: this is just what is defined as the condition of mechanical equilibrium (note, that mechanical equilibrium has nothing in common with the thermodynamical equilibrium: the mechanically equilibrated systems can be very much internally stressed, that is storing within themselves a substantial amount of energy). Let us ask ourselves: how and when can mechanical equilibrium in a morphogenetically active system be reached after a material point has been shifted by a force F? For answer this question we have to explore the situation with greater temporal resolution. The process starts from the moment when a force F is generated (as a result of a mechanochemical transduction) in a point N and stretches a material fibre NM (Fig. 2.3A). That means that a wave of mechanical (elastic) deformation spreads along the direction NM until it reaches M. From this time moment a point M will move towards N with an acceleration (Fig. 2.3B). This will last until similar waves of elastic deformation p, q, r spread from M along the bonds MP, MQ, MR will reach their own fixing points P, Q, R (Fig. 2.3C). From this moment on the vector sum of all the forces applied to the point M will become of a zero value and M will move at a constant rate. Now the crucial question will be to what distance can the fixation points P, Q, R be removed from M. In the other words, what may be the approximate length of continuous mechanical bonds within an embryo? The maximum possible distance should obviously coincide with the entire embryo dimensions, which we may estimate as not exceeding 10^{-3} m. Certainly, for most of the bonds this is an exaggregation; probably they do not exceed the diameter of a single cell, which is in the range of 10^{-5} m. Thus, we get a distance range of two value orders, but most probably fluctuating from 10^{-4} to 10^{-5} m. We may refer now to Forgacs' (1995) estimates, arguing that the elastic deformation wave spreads along the actin fibres at a rate of about 1200 m/s. That means that it should reach the fixing points in 10^{-7} - 10^{-8} s. Such is the time period during which a force is unbalanced and a material point is moving with an acceleration. Odell *et al.* (1981) came to a similar conclusion by claiming that the frequency of elastic oscillations caused by a brief imbalance of the active and resistance forces should be no less than 10^{6}–10^{7} Hz, which is 7–8 orders higher than the reverse rates of morphogenetic movements (see also Oster and Odell, 1984; Goodwin and Trainor, 1985). Is such a small time period enough for the morphogenetic movements to reach their typical rates? Let us suggest, that the motive force is that of actomyosin contraction. Within 10^{-8} s the actin filaments are shifted along each other to an extremely small distance of about 10^{-13} m (Alberts *et al.*, 1989). This is less than 10^{-5} of the total shift during a single actomyosin sliding cycle, but still enough for developing an acceleration of about 3 m/s and reaching a final velocity of about 3×10^{-6} m/s, which already exceeds the typical rates of morphogenetic movements by an order! Certainly, in real situations the minimal displacements produced by the actomyosin system should hardly be less than those corresponding to the entire actomyosin cycle, that is of several micrometers range. That means however, that any morphogenetic displacements are practically immediately balanced by the opposing elastic forces. We come to the

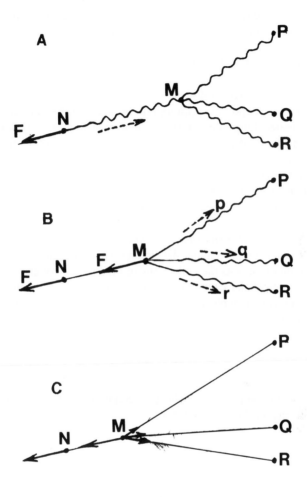

Figure 2.3. A movement towards a mechanical equilibrium of a material point M, which is bound by elastic strings with the points N, P, Q, R and pulled by a force F, applied to a point N. For more details see text.

conclusion that besides some extremely small time periods, morphogenesis goes on under the condition of the mechanical equilibrium. This will be of a great help later on for estimating the relative values of mechanical stresses. On the other hand, one cannot deny that some important and even crucial 'micromechanical' events are taking place within extremely short unbalanced periods: this problem remains completely unexplored up to now, but very promising in the future.

If formulating the above in terms more usual for physiologists, one might say that within the scale of micrometers or so the morphogenetic forces are working in isotonic conditions. And if we enlarge the scale up to millimeters or more, and/or taking into consideration the substantial pauses between the successive rounds of morphogenetic movements, the conditions can be estimated as isometric, because within such a scale the embryonic material can be roughly considered as immobile.

2.1.3. STRESSES, CURVATURES, AND NODULES

In this section we would like to discuss some consequences which follow from classical mechanics and are based upon the presumption of mechanical equilibrium. In spite of most of them being created long ago for simple inorganic bodies, they are quite useful for morphogenesis insofar as they formulate the relations between the stresses and the main geometrical characteristics of a stressed body. Certainly, while taken isolated, these rules are not enough for interpreting any one of the real morphogenetic processes; for doing this the active components should be taken into consideration. However, this cannot be done adequately without first introducing canonical 'passive' rules.

Shells covering pressurized compartments (some consequences of Laplace law).
Many biological samples, including developing embryos, can be regarded from the biomechanical point of view as the collections of pressurized compartments of the various scales (vacuoles, intercellular cavities) surrounded by stretched shells. To these the well known Laplace law can be applied, which relates to each other the internal pressure P in such a cavity, the shell's tension T (produced by this pressure) and the shell's curvature C as

$$P = TC.$$

While for a one-dimensional shell (which is certainly an abstraction) the curvature is inversely proportional to the local radius R

$$C = \frac{1}{R}$$

for real 2-dimensional shells both so called main curvatures (those oriented in two mutually perpendicular directions) $C_1 = 1/R_1$ and $C_2 = 1/R_2$ should be introduced. Now Laplace law takes a form:

$$P = T_1C_1 + T_2C_2 \tag{13}$$

The first and main qualitative conclusion from this law is that under a fixed value of pressure the local tension is inversely proportional to the curvature of a given area: the regions curved the most are tensed the least of all and vice versa. Such a statement can be considered as precise for the symmetrical regions of a shell and as an approximate one for the non-symmetrical regions. However, for biological purposes the approximate view is quite enough. What is even more important, using Laplace law we can pass from a pure statics to some elements of developmental dynamics. Let us assume, for example, that a shell's resistance to the internal pressure is increased along with the increase of the local tension. Under these rather natural asumptions the most curved (less tensed) regions will be deformed by an internal pressure force at a greater rate than the less curved regions, so that even small initial local curvature differences will be aggravated. Another well known mechanical example which may have a number of morphogenetic applications is sometimes called the 'boiled sausage model'.

Consider a pressurized compartment enclosed by a cylindrical shell conjoined at the ends with two hemispheres. It follows from Laplace law that the tension on a hemispherical surface with each of its main curvatures being equal to $1/R$ is

$$T_A = \frac{PR}{2}$$

On the other hand, the tension of a cylindrical surface with the same radius will be

$$T_B = PR$$

which is twice as high as T_A.

Let us estimate now the stretching stresses produced by the internal pressure along a longitudinal and a circular direction (σ_L and σ_C respectively), assuming that the length of the cylinder is much greater than its radius. A longitudinal stretching can be imitated by pushing apart opposite hemispheres. Hence, it is proportional to T_A / t, where t is the shell's thickness. On the other hand, the circumferential stretching is proportional to T_B/t. Consequently

$$\sigma_C = 2\sigma_L \tag{14}$$

This is why an overboiled sausage is always split longitudinally, rather than transversely. More seriously, this is a good illustration of how an overall body shape can affect, in a holistic way, the local stresses. It is easy to see, indeed, that even slight changes in the geometry of the polar regions will be enough for affecting the σ_C / σ_L relations a great deal, even in the areas most removed from the poles. For example, a flattening of a polar region will increase σ_C in the cylindrical region; this is why the gasoline tanks never have flat bottoms. As in the previous example, we may pass from statics towards some elements of a dynamics by assuming that the tissue deformations are stress-dependent. While in the case of a passive deformation proportional to the applied stresses a sausage will simply puff out, approaching a spherical shape, under the assumption of the stress-resistance to increase non-linearly along with the stress increase, a sausage will, in the contrast, elongate itself. And if we propose that the cells tend to be oriented along the main lines of tension, they will take circumferential orientations. This is exactly what is observed in the experiments on cultivating cells onto cylindrical substrates (Levina et al., 1996).

Stretched networks.
There are no embryonic areas (shells, membranes) to be uniformly stretched throughout. Instead, they always consist of single fibers bearing the main tensions and separated from each other by less stressed areas. Any stressed net is a peculiar topological construction

having some morphogenetically important properties. Its main component is a nodule, that is an intersection point of stressed fibers. Even *a priori* we may predict that the most robust nodules should be of order 3, that is just 3 fibers joining each other in one point (for more than 3 fibers to meet each other exactly at the same point is, obviously, a matter of a coincidence).

The stretched networks and nodules can be traced on quite different levels of organization; they can be constructed out of the cytoskeleton or extracellular matrix elements and in most cases out the both together. In such a way the networks become perfectly expressed on the level of large 'cell collectives' and seen, so to say, almost with the naked eye. In any case, however, for each one nodule the equilibrium condition should be fulfilled. This will permit us to distinguish the following situations:

(1) The fibers are curved. That means that the internal pressure forces in the close vicinity of a nodule are comparable with the forces which stretch the fibers themselves, so that the first ones cannot be neglected. That makes it probable that the fibers are stretched just by these pressure forces. The stretching forces are directed in this case along the tangents to the fibers in the nodule point.

(2) The fibers can be approximated by straight lines. That means that the internal pressure forces are negligibly small in comparison with the stretching forces. Let us analyse this situation in more detail:

(2a) One of the nodule angles $\alpha > \pi$. That means that the fiber a opposite to this angle exerts pressure rather than tension (Fig. 2.4A).

(2b) One of the nodule angles $\alpha = \pi$. In this case the tension of one of the fibers (a) is negligibly small in comparison with the both others (Fig. 2.4B).

(2c) Each of the three nodule angles is less than π (Fig. 2.4C). Consequently all of the fibers are substantially stretched. If we know the values of two nodule angles, α and β, and the value A of one of the stretching forces, we can easily calculate the values of the two other stretching forces, X and Y using trigonometrical formulae

$$X = \frac{A \sin (\alpha + \beta - 2\pi)}{\sin(2\pi - \beta)}, \qquad\qquad Y = \frac{A \sin(\pi - \alpha)}{\sin(2\pi - \beta)} \qquad (15)$$

By using these formulae we can 'travel' along a fiber's net from any of its nodules, taking a stress value A as given and then calculating one after another the stress values of all the other fibers, as being related to the stress A. We even can, after travelling around a certain contour, return back to the first nodule and then measure the value of A as a function of all the previous measurements. If this value does not differ much from the initial one we may conclude that the measured contour can indeed be considered as a mechanically closed one; otherwise one have to conclude that it is affected by some outside forces.

An important and easily calculated particular case of equation (15) is associated with the equality of two out of three forces, say X and Y. Now these latter may exemplify the tensions on the apical walls of two neighbouring epithelial cells while A is the tension on the same cells adjacent transversal walls. β is called in this case an *edge angle*. Measurment of the edge angles' values is a good method for estimating, on histological slides, the relations between the apical and transversal cell walls tensions: the greater is β angle the greater also is the apical/transversal tension ratio, and vice versa.

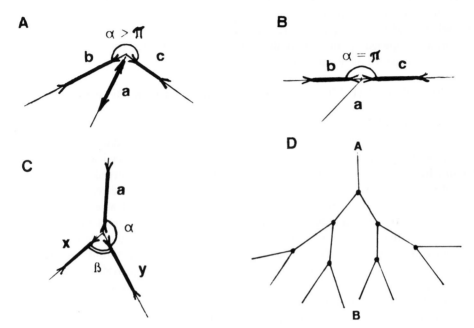

Figure 2.4. Some simplest examples of single nodules and their sets in the mechanically equilibrated stressed nets. For comments see text.

A special kind of nets is associated with the morphogenetically important situation of the *tension gradients*. If the number of tension-bearing fibers is decreased along a certain direction *BA* (they are fused with each other, Fig. 2.4D), the tension values along each of them are correspondingly increased. That is, a tension gradient going down in *AB* direction is established. The same situation can be also interpreted in a reverse sense: if by evaluating independently the similarly oriented tensile stresses in two small areas *A* and *B* of a common mechanically closed net we find that the tension in *A* is definitely greater than in *B* this is enough for claiming that the lines of tension are converged and fused with each other in the direction *BA*.

Tensegrity structures.
This term has been borrowed by biologists from architecture where it was used for describing those constructions consisting of stretched fibers and compressed rods and able to undergo extensive topological transformations without being disrupted. Ingber *et al.* (1994) ascribed them an important role in cell mechanics. The authors suggested that the role of stretched fibers can be played by intermediate or actin filaments, while the role of compressed rods is taken by microtubules. In this view there are the inherent properties of tensegrity structures which can smoothly transform, under a certain tension, a relatively loose and unstable polygonal network into much more firm bundle-like arrangement. The authors believe these kinds of transformations to be among the main tools for providing

holistic cell responses to mechanical forces and energy pumping. Some of these points are disputed by Jones *et al.* (1995). The authors argue, firstly, that a cell is much more sensitive to stretching than to compression (which should not be the case according to the tensegrity model) and, secondly, that the tensegrity structures are too much closed 'in themselves' and hardly capable of reacting adequately to external stimuli. The problem remains to be opened up, once more demonstrating how rudimentary is our knowledge of the real mechanics of the living cell.

2.1.4. RETURNING TO MOLECULAR-SUPRAMOLECULAR DEVICES

Generation of stresses.
Let us recall now some of the text book data for understanding what kinds of supramolecular devices are used for producing the abovementioned kinds of stresses. Starting from *linear contractions* we can easily see that they are provided by a 'classical' mutual sliding of the antiparallel actin filaments (F-actin), connected by the myosin cross-bridges. A shear stress produced by sliding is transformed at a higher structural level into that of a longitudinal contraction owing to the attachement of the peripheral ('+') ends of the filaments to different points of a cell membrane: the portion of the membrane between the both attachement points is that to be shrunk. Along with a classical linear contraction, the same mechanisms can also provide a more or less isotropical (areal) contraction in non-muscle cells. This is because F-actin, depending upon the kind of actin-binding proteins, can form not only bundles of parallel fibers but also a network of the more or less randomly oriented 'actin nodes' denoted as an actin gel (or a cytogel); it is the overall myosin-mediated contraction of a cytogel which gives rise to the areal (roughly isotropic) shrinkage (Fig. 2.5).

Another kind of strains, produced by F-actin fibers, now owed to their polymerization, are *linear extensions,* going at a rate up to about 10 mkm/s (this maximal rate is exhibited during the formation of a so-called acrosomal protrusion in

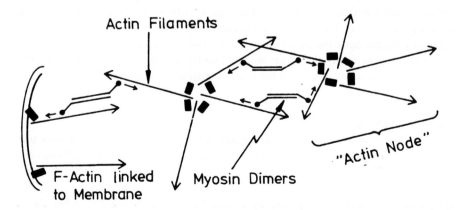

Figure 2.5. Structure of the actin nodes producing in the average an isotropic contraction. To the left the actin - membrane anchoring is shown.

spermatozoa). This process may be responsible for making protrusions in different kinds of morphogenetically active cells. However, the main mechanism for producing linear extensions is associated with the activity of another class of supramolecular devices, the microtubules, which are self-assembled from the tubulin subunits. The microtubules' polymerization is assumed to play a leading role not only in moving apart the poles of the mitotically divided cells but also in the formation and maintenance of the morphological axis in interphase cells (see below for more details).

The morphogenetic role of the *voluminous* stresses and strains (inflations/deflations) is often underestimated, at least by students of animal morphogenesis. These events are, however, of a primary importance in being the main ones opposing the powerful contractile stresses. All of the voluminous stresses are produced by a finely regulated hydrostatic (turgor) pressure within a cytoplasm, vacuolar compartments or intercellular cavities, which exceeds as a rule the osmotic pressure of an external environment.

In all of these cases the excessive osmotic pressure is caused by a local increase in the concentration of some molecules and/or ions (so called osmogenic particles). A 'classical' concept of osmotic pressure demands the presence of a semipermeable membrane, transparent to the water molecules, but not to the osmogenic particles. Another related situation is a 'swelling' of the macromolecules owing due to electrostatic trapping of some water-attracting ions (mainly monovalent cations). According to the modern ideas about the interactions between a solvent and the water clusters (Watterson, 1991) there is no principal difference between a membrane-requiring osmotic pressure and a swelling. The both can produce rather local effects, as for example increasing turgor pressure within single cell protrusions . This involves, probably, depolymerization of a macromolecular F-actin up to G-actin monomeres, increasing hence a concentration of the osmogenic particles (see below for more details). As to the osmotic effects on the entire cells scale, their morphogenetic role is best of all studied in plant species and generalized by the so called Lockhart's (1965) law. This law describes the dependence of a plant cell's growth upon the tension in its wall, the latter determined in its turn by the internal turgor pressure.

According to Lockhart's law the growth rate g of a plant cell (which is, actually, the rate of the deposition of new cellulose subunits onto a cell wall) is proportional to the difference between the actual tension T of the cell wall and a certain basic (threshold) tension S, taking place at a zero growth rate:

$$g = \lambda(T - S) \qquad if\ T > S, \qquad\qquad g = 0 \qquad if\ T < S \qquad\qquad (16)$$

Both S and P may be rather high. For example, in *Avena* coleoptiles the threshold turgor pressure (which corresponds to zero growth) is 6 atm, whereas that providing an optimal growth is about 9 atm (Preston, 1974, pp. 408-414). It is of special interest, that the value of S can itself rise as growth proceeds, thus implying the existence of some feedbacks between growth and the basic osmotic pressure value.

Lockhart's law can be qualified, in our view, as an important generalization, applicable far beyond the special problems of plant growth. Its main advantage is in linking the active stress-relaxing response (growth rate) g with the momentary stress value $(T - S)$, the latter being, in its turn, a complicated. function of cell metabolism, averaged within a finite time period. In this way, Lockhart's law implies an active memory of a developing system about its finite past.

Addressing now the osmotic events in developing animals, we should refer first of all to the turgor pressure within intercellular cavities and the vacuoles. This pressure often shows periodic, in many cases of several minutes, oscillations (Beloussov *et al.,* 1993). Those implying periodic vacuolar swelling will be considered in greater detail somewhat later (see 2.3.7), as being associated with a so called pulsatorial growth and morphogenesis. As to intercellular cavities, we would like to point out first of all that their pressure values are rather substantial: for example, in amphibian embryos the turgor pressure within a blastocoel is estimated to be about 325 mosm = 70 N/cm^2 = 7 atm. If attempting to derive from these data the value of a stretching stress within a blastocoel roof we come to the abovementioned (p.76) controversy. Namely, if suggesting that the stretching force is spread uniformly throughout an entire cross section of a blastocoel roof we obtain a not too high value of 2×10^6 N/m^2 (see for discussion Beloussov *et al.,* 1994). If assuming however (what is closer to the reality) that the entire tension is beared by a sole submembraneous microfilament's layer of 0.1 mkm thickness the stress value will rise up to $\sim 10^9$ N, as in our previous, more speculative estimations. Probably, it would be reasonable to become accustomed to these orders stress values!

Similar turgor pressure values are established in the subgerminal cavity of an early chicken embryo and within a mammalian blastocyst. In both the latter cases the pressure exhibits regular pulsations of a few minutes' periods. In all cases the intracavital excessive osmotic pressure is generated owing to a polar (apico-basal) asymmetry in the arrangement of ionic pumps and channels of the covering epithelial cells: namely, the channels are located mostly within the apical membrane domain, while the pumps (Na^+, K^+ ATPase) in the latero-medial domain, both domains being separated from each other along the cell membrane by impermeable tight junctions. Because of this, a vectorized apico-basal transport within the cavities of at least a sodium, together with the accompanying anions, is maintained (Stern, 1984).

The turgor pressure in embryonic cavities is indispensable for morphogenesis. Let us mention just two examples, related to quite different taxonomic groups: a bud of a fresh water hydra stops growing when the turgor pressure in the gastral cavity is reduced (Wanek *et al.,* 1980) and the eye development in a chicken embryo shows numerous anomalies if the turgor pressure within the eye chamber is diminished (Coulombre, 1956).

Transmission and stabilization of stresses.
As we shall see later on, the stresses produced even by single cells can be transmitted via the neighbouring cells or extracellular matrix to distances greatly exceeding cell diameters. This is true most of all for tensile stresses, while those of a directed (microtubules-mediated) pressure are damped much more rapidly. The main structures which bear and transmit tensile stresses are the intermediate and actin filaments, integrin molecules, cell adhesion factors, and the extracellular matrix. Almost always the transmission of tensile stresses is far from being a passive process going on along stable preexistent conductors. Instead, the very assembly, orientation, and subsequent disassembly of all the tension-transmitting structures is itself stress-dependent. Under these

conditions a stress transmission implies an entire hierarchy of feedbacks, and can be only arbitrarily distinguished from the stress generation proper.

In relation to the stress-transmitting properties of the various fibrous nets the so called percolation theory is of interest. As proved by topology the nets destined to transmit stresses, or some other messages, from one space point to another should have a considerable transmission redundancy: they can be, up to a certain threshold, disrupted in many points without losing their continuity. Meanwhile, at a certain moment of disruption (called the percolation threshold) the continuity should be lost, and a net dissociated into several unconnected domains. There are some evidences that in developing organisms the stressed nets of the different structural levels are slightly above the percolation threshold (Forgacs, 1995). That permits them, on the one hand, to be integrated over large scale domains, and, on the other hand, to be anisotropically oriented and to distinguish the signals coming from different directions (a highly redundant net would be much more isotropic and insensitive to a directionality).

The role of a cell membrane in the transmission and stabilization of stresses deserves special consideration. Taken in isolation, the membrane (lipid bilayer) cannot bear any substantial stresses, since the breaking tensions correspond, for erythrocytes to 2-4% (Evans and Skalak, 1980), and for osteoblasts even to 0.3% (Jones *et al.*, 1995) of their planar stretching. Under these circumstances the possibility of a cell surface being stretched up to 100% and more is related, as a rule, to the unfolding of the membrane, previously extensively folded. The foldings–unfoldings themselves are also quite far from being purely passive phenomena. Membranes of many kinds of slightly stretched cells have been found to flicker perpetually at amplitudes of about 300 nm and at frequencies of 1-25 Hz (Levin and Korenstein, 1991). Much more extensive and slow waves are exhibited by a number of cleaving eggs (for example, in Nematodes) in the interphase period. On the other hand, when tension increases (in the case of egg cells, just before cell division), the undulations decrase and the membrane becomes very smooth. This may be qualified as a kind of common membrane/underlaid cytoskeleton phase transition from a more noisy quasi-Brownian state to a much less noisy one, the cytoskeleton playing a role of a quieting, or a rectifying element (see Jones *et al.*, 1995 and Peskin *et al.*, 1993, for discussion about 'Brownian ratchets'). It seems probable that while the membrane is quieted a cell becomes much more sensitive to the mechanical impulses as well as probably to some others.

A cell membrane's dynamics is also of primary importance for fixing and stabilizing the achieved stresses. This is associated with either endocytosis (a resorbtion and internalisation of membrane components) or exocytosis (insertion of new portions within a cell membrane). It is the endocytosis which stabilizes the contracted cell state: a contracted part of a cell membrane is firstly ruffled and then, because of endocytosis, engulfed (sometimes endocytosis is replaced by a shedding, that is a rejection of membrane parts towards the outside). On the other hand, the extension of an entire cell or some of its parts, in order to become stable, should be followed by the insertion of the additional portions of cell membrane, which is achieved by exocytosis. An irreversible fixation of either extension or contraction strains by insertion/resorption of membrane subunits may be characterized as a plastic (non-elastic) component of a cell response to a stress; it usually comes into action in a few minutes after the elastic component of the reaction.

2.1.5. A STRESS-DEPENDENCE OF SUPRAMOLECULAR DEVICES

By the modern views, virtually all the types of cells, rather than   only some highly
specialized ones, contain mechanosensitive elements. The most probable candidates for the
primary mechanosensors are:
   – stretch-activated ionic channels for Ca^{2+}, Na^+, K^+ and H^+ (Banes et al., 1995);
   – transmembrane proteins, including integrings, G-proteins and the entire focal
adhesion complexes (Burridge and Chrzanowska-Wodnicka, 1996);
   – elements of a cytoskeleton with the associated peripheral proteins, for example
phospholipase C, a key enzyme for rising free intercellular $[Ca^{2+}]$ (Jones et al., 1995).
   In a number of cases a primary response to a stretch can be traced already within
milliseconds and   expressed in the rise of free intercellular $[Ca^{2+}]$ and the shifts of a
transmembrane potential (Banes et al., 1995). As suggested by Burridge and
Chrzanowska-Wodnicka (1996) and by Shyy and Chien (1997) one of the most important
pathways starts from the β-integrin subunits which as responding to mechanical stresses
are clustered into focal junctions. A next step will be the activation of the tyrosine kinase
receptors associated with the focal junctions (so called focal adhesion kinases, or FAKs).
Activated FAKs affect in their turn the GTP-binding proteins from Rho subfamily. These
latter perform numerous functions. One one hand, they are the starting points for either a
mitogenic cascade or that leading on the contrary to a cell death (apoptosis). On the other
hand, they promote a number of membrane- and cytoskeleton-bound activities, including
the assembly and contraction of actomyosin filaments and thus the generation of an
intracellular tension (as a response to the external one). That may underlie a
morphogenetically important stretch-contraction cell–cell relay, to be numerously
mentioned later on. In any case, an entire cell including its nucleoplasm should be regarded
as an integrated mechanochemical system, its mechanical links being coupled in many
points with the chemical transformations (Maniotis et al., 1997). It is also worth
mentioning that some of the DNA loci are themselves mechanosensitive (Resnick et al.,
1993). Among this bulk of rapidly accumulating analytical evidences a structural basis for
the various morphogenetically important stress-mediated feedbacks will be of an utmost
interest for us. In such an aspect we will review now in greater detail a stress sensitivity of
the different elements of a cytoskeleton and a cell membrane.

Actin structures.
Among the most instructive in this respect are the experiments performed on the
cytoplasmic strands of a plasmodium of *Physarum* fungi (Fleischner and Wohlfarth-
Bottermann, 1975; Wohlfarth-Bottermann, 1987). The authors observed that a 5 sec
stretch of a strand was enough for initiating a rapid polymerization of actin which could
balance the applied force (up to 10^3 N/m) and to transmit it thus along the entire strand.
The actin fibers were kept intact only under the isometrically contracted state; both under
isometric relaxation and isotonic contraction they lost the fibrillar structure and were
depolymerised. This observation is of a particular interest, because, owing to the low rate
of morphogenetic movements the mechanical state of embryonic tissues is close to
isometric; we can see now that this promotes the maintenance of the fibrillar actin state.

A cytoplasmic strand also exhibits a series of the regular 2 min autonomous pulsations, their amplitude increasing after application of a stretch stimulus; these pulsations point hence to the existence of stretch–relaxation feedbacks.

A stretch-dependence of actin polymerization has also been observed on several animal samples, namely epidermal fish cells (Kolega, 1986) and in *Xenopus laevis* early gastrula tissues (Beloussov *et al.*, 1988). In the latter case 1.4 – 1.2–fold stretch was enough for producing in several minutes pronounced microfilament bundles oriented in the stretching direction together with the associated focal adhesion sites.

Microtubules

These are also mechanosensitive. As argued by Skibben and Salmon (1994) the kinetochore microtubules from a mitotic spindle add to themselves tubulin subunits under tension and remove subunits when being compressed. Besides, a microtubules' self-assembly (or, better to say, a self-organization, since the arised patterns require energy input and exhibit obvious properties of the dissipative structures) are largely dependent upon the geometry of the immediate environment (Nédélec et al., 1997).

Very interesting responses of the growing axons to stretching and relaxation (Dennerly *et al.*, 1989) are also probably mediated by microtubules. The authors found that after exceeding a certain threshold of stretching force (100 microdynes) the axon continues to elongate itself actively, lowering the externally applied tension. On the other hand, after a sudden diminution of tension an axon contracts, thus doubling sometimes the pre-experimental steady-state tension. That corresponds to the assumption of a stress-mediated polymerization – depolymerization of the microtubules within the axon. Along with that, the processes of membrane insertion-resorbtion may be involved.

Stress-dependence of the resorption–insertion processes within a cell membrane (a cell wall)

Such a dependence is best of all studied in growing plant samples. Let us discuss some of the situations associated with the above described Lockhart's law.

So far as a cell growth (= cell wall growth) is, on one hand, stress-inducible and, on the other hand, leads to the relaxation of a tensile stress within a wall, the following situations are possible:

(1) a growth-mediated relaxation (dR) is less than the rise of the internal turgor pressure stress (dP): $dR < dP$. As a result the tension on the surface will be increased.

(2) in the contrary, $dR > dP$; that means the reduction of a net cell wall tension.

(3) a boundary case corresponds to the relation $dR = dP$. That means that during a growth process a cell wall tension remains the same, or, in the other words, that the entire work of a turgor pressure is spent on the cell wall's extension. Such a situation has the advantage of being stationary. In a so-called stress theory of bacteria and fungi growth (Koch and coworkers, cited from Harold, 1990) it is expressed by the equation

$$PdV = TdA \qquad (17)$$

where P is the osmotically-driven turgor pressure, dV is the volume increase of the cell under this pressure's action, T is the tension in the surface (the stretching force divided by cell wall thickness) and dA is the increase in cell surface area. Under these assumptions the tension on the surface will be linearly proportional to the turgor pressure. From this it follows that the simplest shapes which keep a constant surface tension value during the entire growth period are the cylinders (filaments) of constant radii, as keeping a linear proportionality between a volume and a surface area. This is probably the reason for a wide spread of such shapes among bacteria and fungi.

On the other hand, a decrease of the tension on the surface, associated with a hyperproduction of surface material subunits seems not to be so dangerous for an organism as an increase in tension. A natural consequence of such hyperproduction would be the folding of a cell wall. Since, as mentioned before, in cylindrical bodies that are under internal pressure the circular tensions are twice as great as the longitudinal ones, it seems most probable of all that an overproduction (and hence a folding) of surface material will occur in just the circular directions. In such a way a circular cytotomy of a bacterial cell may be initiated. And in a more general context (see below) such a hyperreaction to a mechanical stress should be regarded as one of the first rudiments of a real shape complication, that is of a morphogenesis.

There is a substantial amount of data indicating a similar stretch-induced insertion of a new material into the walls of animal cells. In our experiments with stretching the pieces of *Xenopus laevis* embryonic tissues (Beloussov *et al.*, 1988; Beloussov and Luchinskaia, 1995) the cells produce extensive protrusions oriented just in the stretching direction as soon as in 10-40 min after the beginning of stretching; the protrusions continue to grow well after the external stretching has stopped. This is accompanied by the enlargement of those cell–cell contact areas to which the stretching forces are applied (Fig. 2.6, cf. A and B). Each of these events requires the production of new portions of cell membrane.

The stimulation of exocytosis (and hence area increase) in lung tissue after being stretched by 5-25 percents in area has been demonstrated by Wirtz and Dobbs (1990); this was associated with an increase of the free Ca^{2+} concentration. Recently, Fink and Cooper (1996) reported that in the epithelial cell membranes of a fish's (*Fundulus heteroclitus*) embryos the activity of the apical membrane turnover is positively correlated with the tension, either applied artificially or generated by embryonic tissues themselves at the leading edges of the marginal cells.

On the other hand, several sets of observations and experiments indicate that the relaxation of a cell surface is followed by its resorbtion, due either to endocytosis, or to shedding, or to the both together. In Fig. 2.6 C a cell shown by a dense arrow is at least relaxed (if not slightly compressed) by the actively extended cells to the left of it (small arrow). As a result, the first cell's membrane is extensively folded to be later on resorbed by shedding. Remarkable is an abrupt (threshold) transition from the extended to the folded state of a cell membrane. Later on we shall return to this typical situation. A fascinating example of a relaxation- induced membrane dynamics is given by the zygote of a Teleostei fish, *Misgurnus fossilus* (Ivanenkov *et al.*, 1990). By sucking off a part of its yolk it can be transformed into an amorphous flattened mass; in less than a dozen of minutes, meanwhile, it restores, owing to the volume diminishment, a tensed spherical shape. This is achieved partly because of the internalization of a substantial part of its membrane and partly because of the creation onto it of a peculiar stressed network (Fig. 2.6.D), which will remind us soon of the stressed networks in multicellular collectives (cf

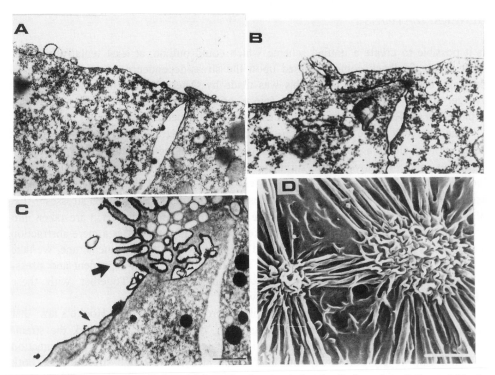

Figure 2.6. Some examples of the stretch- and relaxation-induced cell dynamics. A-C: amphibian's (*Xenopus laevis*) embryo at the early gastrula stage. A: non-stretched ectodermal cells with only few small contact areas. B: same cells cells after 10 min duration horizontally oriented stretch. Cells create protrusions in the direction of stretching, and cell-cell contacts are increased as compared to A. C: an abrupt transition from a raw of extended cells (small arrow) to a cell with an extensively folded membrane, undergoing resorption (dense arrow). D: a membrane of a *Misgurnus. fossilus* zygote within several dozens minutes after extensive relaxation, revealing a stressed network pattern. A-C original, D from Ivanenkov *et al.* (1990), with the authors permission.

Fig. 2.16 and the corresponding comments). Also, in the yolk syncytium of another fish embryo, *Fundulus heteroclitus*, endocytotic events have been observed just within the mostly folded and compressed regions of the periblast (Betchaku and Trinkaus, 1986).

A somewhat similar restoration of a spherical shape with diminished volume has been observed in sea urchin blastulae after immersing them in a hypertonic medium (Beloussov and Zhadan, 1993). Within the first few minutes the blastulae created numerous irregular folds which, however, soon disappeared so that the samples became spherical again, with their blastocoel volumes decreasing 1.8 – 2-fold. Accordingly, the total surface of the apical cell walls also diminished in 1.2 – 1.4 times. At the same time the average cell volume increased by 1.5 – 2 times (obviously because of the water flow out of the shrunk blastocoel) and their transversal walls areas by 1.8 – 3 times. That points to the resorption of the cell surface material at the relaxed (as a result of hypertonic shrinkage) apical cell walls and insertion of the material into the stretched transversal walls.

2.1.6. THE CONCEPT OF AN ACTIVE CYTOGEL, WITH SOME ADDITIONS AND
GENERALIZATIONS

Is it possible to create a unified scheme which could outline, at least tentatively, some
most simple feedback contours based upon the stress-dependence of mechanochemical
devices? The first attempt to do this was made by Oster *et al.* (1985) in their 'active
cytogel' model. Cytogel means here a 'piece' of an active (energized) actin gel. Let us
explore this model, taking the liberty of extending and generalizing it.

The main idea of the model is presented as a standard stress-strain graph, the strain ε
being plotted along a horizontal axis (a positive strain corresponds to a stretching–
extension, and a negative one to shrinkage–contraction) while stress σ is plotted along
the vertical axis. The only one kind of stress which produces the volume's increase is
associated with the turgor pressure (σ_{tp}). Besides, two different contraction stresses – that
of a passive elasticity (σ_{pe}) and of an active actomyosin contractility (σ_{ac}) are taken into
consideration. However, insofar as a 'naked piece of an actin gel' is a mere abstraction
and in reality it is always surrounded by and closely linked to the cell membrane, we take
the liberty of adding to the model the two abovementioned strain-dependent and stress-
producing membrane-bound activities: exo- and endocytosis. Together with these
additions the entire picture will look as follows.

The passive elasticity is assumed (as a first approximation) to obey Hooke's law, that
is, to show a linear strain-dependence (Fig. 2.7A, σ_{pe}). On the other hand, the strain-
dependence of an active contractility should be treated already in the first approximation
as non-linear, steeply decreasing both at low and at extremely high stretching (in both
cases as a result of F-actin depolymerization) – Fig. 2.7A, σ_{ac}. As to the turgor pressure,
it should be, firstly, a function of the osmotic events and, secondly, of the sample's
compression. We suggest that it takes the greatest values at a low stretching, because
of the relaxation-induced F-actin depolymerization and, hence, an increased concentration
of the actin monomers (Fig. 2.7A, σ_{tp}). The compression of a sample (if the latter is
possible) should also reinforce the hydrostatic pressure. Probably, a second osmotic
pressure maximum (not shown) should be expected to occur at high extension, again as a
result of the actin depolymerization now caused by the over-stretching. However, any
strain-dependent modulations of the turgor pressure will not affect very much the overall
picture: the main role of the turgor pressure is in roughly counterbalancing the
contraction-generated stresses.

Exocytosis itself is directed towards relaxation of the stretching stresses while
endocytosis, in the contrast, tends to increase them: as we have learned from the
previous section, the exocytosis is stimulated by the stretching of the sample, while
endocytosis, on the contrary, is stimulated by its relaxation or compression. Taken
together, these two opposite activities (σ_{ex} and σ_{end}) may look as if they compensate each
other, and thus are of no importance. We hope to show, however, that if we take into
consideration the T_{ch} -s of the processes involved, this is far from being the case.

What can we deduce from such a deliberately rough and tentative construction? The
answer can be given in two successive approximations. In the first we will not take into
consideration the differences in the T_{ch} -s. For the next order approximation, however,

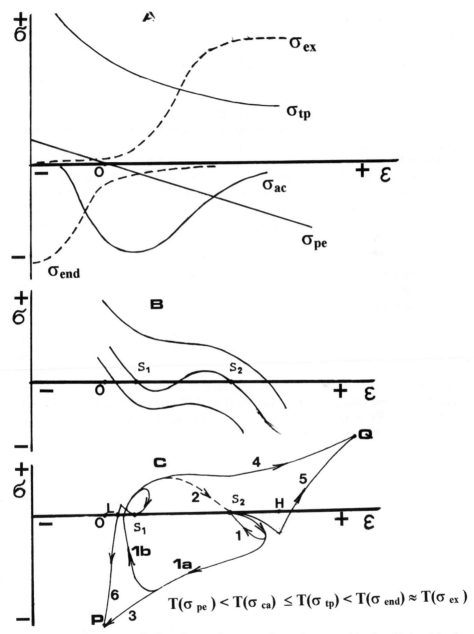

Figure 2.7. Models of stress - strain dependences for a cytogel+membrane combination. Horizontal axis: stresses (positive are to the right and negative to the left of a point O). Vertical axis: strains. Positive ones correspond to a sample's extension while negative to its shrinkage (contraction). Frame A gives a rough diagramm of the forces involved, B illustrates a potential bistability of a sample and C shows some consequences of the T_{ch}-s differences. For more comments see text.

these differences will become crucial.

So as a first approximation we simply summarise all of the superimposed stress-strain plots. The main conclusion will be that the resulting curve takes on an S-shape, crossing the horizontal axis either in one or three points (Fig. 2.7B). That means that a sample with such cytomechanical characteristics will have either one stable length (resp. volume), which corresponds either to a relatively low or to a relatively high strain value, or two such lengthes (volumes) S_1 and S_2, separated by an unstable interval. (Note that the mechanical equilibrium, as discussed in 2.1.2, occurs at *any* point of the phase space, rather than at the stable points only, since the time required for establishing it is many orders smaller than the T_{ch}-s of the relaxations now considered). As a result we tentatively conclude that the system under study is potentially bistable.

Meanwhile, the main advantages of the model will become obvious if we accept (which is really taking place) that the T_{ch} -s of the variables described (the rates of their responses to external pertubations) differ considerably from each other. The smallest T_{ch} belongs, certainly, to the passive elastic stress, which reaches its full strength immediately after the breaking of the mechanical bonds. As for an active contraction stress, however, it requires no less than a few seconds to achieve its maximum. For the osmotically-driven turgor stress this time will lie in the range of dozens of seconds while for the exo- and endocytotic events it will lie in a range of no less than several minutes.

The consequences of such a 'delay hierarchy' are illustrated by Fig. 2.7C. Suggest that a sample has two stable points S_1 and S_2 and is located initially at the right (higher strain) point S_2 (it is worth mentioning, that our way of reasoning holds true also in the case of a single stable point). If we shift now a sample by an external force further towards the right (to even higher positive strain values) σ_{pe} value will be increased. By doing this quite gently and slowly (that is, giving enough time for all of the stresses to reach their stationary values) we shall bring the sample back to S_2 where it will rest for an indefinite period; same will take place in S_1 vicinity (Fig. 2.7C, small cycles 1 and 4). Such cases corresponds to a highly viscous state. Now assume that the sample is released very rapidly (as if cutting off the elastic bonds with its surroundings). Under these conditions the force of elastic contraction, without being opposed by much more slowly developing osmotic forces (but probably supported at a proper time moment by the forces of active contraction) will shift the sample's state very much down and to the left along a route *1a*, perhaps even overshooting S_1. What happens later will depend, most of all, upon the relations between osmotic forces and endocytotic activity. If the former dominate, the sample, after some considerable contraction, will increase its turgor pressure and hence become extended. The sample will then either come along *1b* route to S_1 and stay there, or, in the presence of powerful, although if slowly developed, osmotic force will be shifted backwards to S_2 along a route *2*. the next small perturbation will induce a new similar cycle, etc.. That gives a rough idea of so called osmocontractile pulsations, to be treated in more detail in 2.3.7. If, however, the endocytosis takes the major role an entire contracted surface can be rapidly resorbed up to a practically zero value and the sample will be irreversibly 'frozen' in a highly tensed state, far removed from both of the initial equilibrium values (a trajectory 3 up to a stationary point P). A similar, although oppositely directed response will take place in the case of an extensive stretch-mediated exocytosis (trajectory 4, up to Q). Finally, if a sample is arrested by an external force for a long enough time period either at an abnormally high or low strain (points H and L respectively) it may exhibit either exocytosis-mediated extension (trajectory 5, up to Q) or endocytosis-mediated contraction (trajectory 6, up to P).

In any case, we can see that the stationary values S_1 and S_2 are a kind of abstraction (although theoretically useful), since they are stable for only rather small and special stress/strain deviations. In more robust cases a system exerts either oscillations, or some aperiodic and practically irreversible shifts, which largely exceed the initial perturbations, and are associated, as a rule, with a change of a sign of the stress (forced extension is followed by contraction, and vice versa). The self-organizational capacities of even such a deliberately simplified system are actually enormous! And to a great extent they rely on T_{ch} -s differences between its variables, thus supporting Polezhayev and Ptytsin idea (see p. 62) that these differences may become a proper way for complicating a system's structure. From an even more general viewpoint, systems of such a kind are of interest by smoothing out to a great extent the temporal differences between the dynamical variables and the parameters: what we see instead is a conglomeration of the variables with quite similar and fluctuating characteristic times. This is but a natural development of a tendency which we have already discussed in relation to autooscillating regimes (see Fig. 1.19 and the corresponding comments). The developing organisms seem to be very non-pedantic in their attitude to the formal categories.

2.1.7. A CYTOGEL PLUS CURVATURE: ACETABULARIA MODELS

The next natural step after introducing membrane bound factors would be to take into consideration an overall curvature of the membrane or, more generally, a curvature of a sample's surface. Several models employing this factor within a "chemodiffusional context" have been suggested recently (Spirov, 1989; Cummings, 1994). However, some important papers developing these ideas appeared even earlier. They were dealing with a giant unicellular algae, *Acetabularia*, which can be considered as an enormous piece of actin gel covered by a cellulose layer of a definite curvature. Goodwin and Trainor (1985) interpreted the *Acetabularia* morphogenesis in the following way.

The main actors in the suggested scenario are the osmotically-driven turgor pressure which inflates the tip, free Ca^{2+} ions, and a cellulose surface. The central authors' idea is that an increase of $[Ca^{2+}]$ within a cytoplasm affects in a non-linear way the elastic modulus (and, thus, the deformability) of the cortical layer of actin gel. On the other hand, $[Ca^{2+}]$ increase is assumed to be proportional to the tension on the cellulose surface (probably because of the tension sensitivity of Ca^{2+} channels). The tension is, in its turn, according to a Laplace law, inversely proportional to the local surface curvature. Now, the suggested sequence of events may be as follows. At the beginning $[Ca^{2+}]$ is the highest at the very tip, thus promoting the tip stretching owing to the increased deformability of its surface. As it is stretched the tip increases its curvature, thus reducing the tension. As a result $[Ca^{2+}]$ at the tip will be decreased and the maximal $[Ca^{2+}]$ will be now shifted to the proximal. Here it will create a concentration ring, which appears to be unstable for small $[Ca^{2+}]$ perturbations. Thus it is split into several sectors of a higher $[Ca^{2+}]$ which correspond to the lobes of the *Acetabularia* whorl. In such a way the successive shapes of the *Acetabularia* tip can be linked with each other via the curvature-mediated Ca^{2+}

dynamics. What we have here is, hence, a closed '+,-' feedback contour, one of its links being chemical (Ca^{2+}-dependent), while another is geometrical (the role of the curvature). The modest temporal differences between the variables play here no less important a role as in the previous constructions; in the absence of a temporal delay between the $[Ca^{2+}]$ and the curvature changes the whole process will be immediately arrested.

It follows meanwhile from the already cited Martynov's (1982) paper (p.64), that for some events of *Acetabularia* morphogenesis at least, even a definite $[Ca^{2+}]$ prepattern seems not to be necessary. According to equation *(8)*, the formation of a whorl can be indeed interpreted as a direct result of a loss of mechanical stability under the influence of a homogeneously distributed turgor pressure (see Fig. 1.28c). Worth mentioning that in this construction the idea of the temporal delays between the different variables has already been employed in its full scale. Along with that, the usage of the curvature as a link in the morphogenetic feedbacks reminds us once again that in the real developing systems the T_{ch}-s differences between the dynamical variables and the parameters may be small to the extent which makes it difficult their discrimination from each other. Is indeed the curvature a parameter or a dynamic variable? Until it is kept constant, it should be attributed to the important developmental parameters, as being unambiguously linked with the tension, or with what Martynov calls the 'constructive rigidity' of the rudiment's surface (for example, the flat surfaces can be much more easily compressed than the convex ones). Meanwhile, as soon as the curvature starts to change itself it becomes just a dynamic variable which we are tracing. In addition to this, a system which is complicating its curvature (produces new minima and maxima of it, separated from each other both in space and in time) increases, at the same time, its 'dynamical dimensionality', that is, a number of its dynamic variables: such a situation reminds that explored by Kaneko and Yomo (1997), see p. 56. The cases of a dynamical, or opened dimensionality look as typical enough for the developmental processes, and morphogenetically productive. Hence, they require more extensive modelling and experimental studies. We shall become confronted with the similar situation when considering again the curvature associated problems as applied to multicellular embryos (2.3.7; 2.5.1).

2.2. Dynamic structures involved in single cells activities.

2.2.1.POLARIZATION OF CELLS

Almost all the cells, both in embryos and in adult organisms, spend the main part of their life cycles in a polarized state, having a symmetry order no greater than *1·m*, (if not at all *1*). Meanwhile, at a certain moment of their history they come to this state from a more symmetrical, and even from a spherical, shape. The best known example (already briefly discussed in Chapter 1) is the polarizarion of an egg cell (an oocyte). With this exception, cell polarization is in many cases reversible and even transitory. For example, most of the proliferating epithelial cells take a spherical shape before each next division for returning again, after the completion of a cell division, into a polarized one. A similar succession of events can be imitated in an accelerated way if experimentally relaxing the tangential tensions in a cell layer: a relaxationally induced cell sphericalization occurs under these conditions within a few seconds, while the restoration of a polarized state takes some few minutes (Beloussov *et al.*, 1990). The cells of a mesenchymal origin, and among them the extensively studied fibroblast cells, also may become morphologically polarized and depolarized again within a similar time scale, by depending mostly upon their interactions with a substrate or with other cells. A monopolar state can be reversibly exchanged with a bi- or tripolar one; the polar end, marked by a cytoplasmic protrusion or a lamella, can also rotate itself all around the cell's periphery (Alt and Kaiser, 1994). Therefore a polarized state is accompanied by a whole family of the related ones.

As we have already discussed in Chapter 1, a polarization of a body may be either 'imprinted' from a macroscopic dissimmetrizer (according to Curie principle), or emerge 'de novo', that is, from minor fluctuations, at the conditions of a loss of stability of a non-polarized state. Most probably, the cells which are ready to become polarized (and, first of all, the egg cells) are just on the verge of stability, being thus satisfied in most cases by rather weak external dissimmetrizers, which trigger an entire cascade of the internal events. As shown in one of the best studied examples of a cell polarization, namely in the zygotes of brown alga, *Fucus* and *Pelvetia* (review: Kropf, 1994), such a cascade implies: (1) concentration of F-actin on the presumptive rhizoid pole; (2) establishment of microtubules arrays which probably rotate the nuclei and the centrosome, orienting centrioles parallel to the polar axis and providing connections between the nuclear region and F-actin zone; (3) spatial segregation (and/or local activation) of ionic channels and pumps which cause Ca^{2+} and H^+ influx to a rhizoidal pole of a zygote and their outflux from the equatorial region (Fig. 2.8 A,B). All of these events obviously support each other: F-actin anchors ionic channels, while local acidification stimulates actin polymerization.

The feedbacks mentioned are probably mediated by tensile stresses in the cortical actin network of a polarizing cell. As suggested by Harold (1990), the role of light in polarizing a *Fucus* zygote may just be in disrupting this network locally, promoting its concentration onto a shadowed pole, which later on gives rise to a rhizoid (Fig. 2.8 C). This is in agreement with the data indicating the formation of bi- and multipolar *Fucus* eggs under the moderate dozes of a cytochalasine, able to produce similar disruptions (Isaeva, 1984). In any case, the mechanisms suggested are among those being effective just on the verge of stability of a non-polarized state.

Although the polarizing mechanisms in the various animals zygots are far from being exhaustively studied, there are good reasons to believe that they do not differ very much from those above described. In all the known cases they involve a local concentration of a subcortical actin and the associated clusterization of the membrane components (see Chapter 3 for more details). The segregation of ionic pumps and channels and, hence, the establishment of a vectorized ionic transport also seems to be a universal phenomenon (Jaffe, 1981). The main conclusion would be that in order to establish and maintain a polarized structure, the living system should be in a state of a flux (and, hence, removed from a thermodynamical equilibrium). As we shall see soon, quite the same is true for the movement and a division of a cell body.

The thermodynamical non-equilibricity in no way excludes (and even implies) a general trend towards a local equilibrium. In this respect we would like to refer again to Svetina and Zeks (1990) model, which successfully imitated the polarization-associated changes

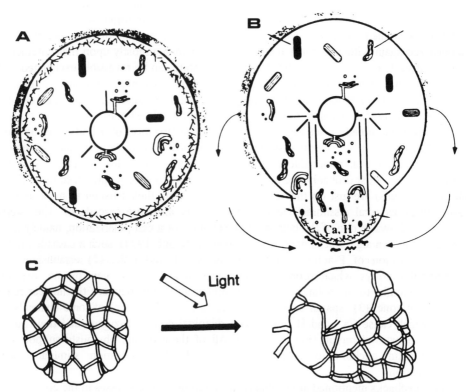

Figure 2.8. Polarization in the *Fucus* zygote. A: a newly fertilized egg without an inherent polarity. B: a germinated polarized zygote with many of its components asymmetrically distributed and ionic flows coming out of a thallus into the rhizoid tip. C: a 'photomechanical' model of a zygote polarization implies that the light can cut some of the cytoskeletal threads causing the remainder to migrate to a rhizoid pole and form a cap. A, B from Kropf *et al.*, 1990, C from Harold, 1990, with the author's permission.

of a zygote shape by assuming that its membrane had been moved towards a minimal bending energy value. Let us stress once again, that the membrane is assumed to come to a certain local (not an 'absolute') energy minimum, and that such a tendency would be impossible if a sample was not previously to a great extent energized.

As we have already told before, not only the oocytes but almost all of the embryonic cells are more or less stably polarized. A perfect polarity, which is in some cases even 3-dimensional, is exhibited by the epithelial cells. Its main component is the apico-basal polarity, based obviously upon the similar segregation of the ion channels and pumps. In embryonic development it is established as soon as the first rudiments of a primary body cavity, the blastocoel, are formed, this taking place sometimes already at the two blastomeres stage. From this stage on the apico-basal epithelial polarity is strictly inherited by the subsequent cell generations.

The only known methods of reversing this kind of polarity have been described by Stern and MacKenzie (1983): it was in affecting a chicken's embryonic epithelium by an external electrical field, its positive pole directed towards the apical surface of epithelial cells. Within about half an hour an entire intercellular architecture, including the rings of microfilaments, positions of cell nuclei, and the sites of extracellular matrix deposition, has been completely reversed in the apico-basal direction. The authors suggest that the primary event was the 'lateral electrophoresis' of the integral proteins, and most probably the ion pumps, within a plane of a cell membrane: the pump proteins were shifted towards anode, which in these experiments have been oriented apically. In any case, such a rapid redistribution of practically all of the cell components indicates that the stability of the apico-basal epithelial polarity is in no way identical with its rigidity: again, it looks as being a highly dynamical one.

Polarization is, certainly, far from being a last step of a cell's dissymmetrization. In egg cells it is followed, as a rule, by the determination of a dorso-ventrality (see Chapter 3 for more details), which is linked, in its turn, with the still enigmatic process of a right–left dissymmetrization. The epithelial cells, in addition to the apico-basal polarity, exhibit also a so-called disto-proximal one, which is oriented as a rule perpendicularly to their apico-basal axes coinciding with a longitudinal (disto-proximal, or antero-posterior) axis of an entire animal's body. Such a polarity is quite well seen, for example, in the insects hypoderm (Lawrence, 1992) and can be also revealed by indirect methods in the ectoderm of hydroid polypes (Labas et al., 1987) and other species. This kind of polarity gives a remarkable example of an intimate linkage of the structural events revealed at two very much removed levels – that of the single cells and that of a whole organism. Unfortunately, very few is known about the determination and the structural basis of the disto-proximal cell polarity. Most probably, it is also stabilized at the membrane level, since it involves, in the case of hydroid polyps, a redistribution of ion channels and, in the case of insects, changes in cell adhesion properties.

2.2.2. CELL CRAWLING AND EXERTION OF TRACTION FORCES

What should be discussed briefly in this section is usually defined in text-books as the mechanisms of cell crawling. It has turned out, however (and this is of a primary morphogenetic importance) that during the process denoted by this term some main types of tissue cells (epithelial cells and, most of all, fibroblasts) are not so much crawl, that is change their positions relative to their neighbours, as produce pulling forces, applied to an extracellular material. "In other words, motile cells can sometimes act more like conveyor belts or winches, rather than like tractors" (Harris, 1994). Consequently, their main function seems to be in establishing and maintaining tensile fields over macroscopic distances. We shall return to this important aspect somewhat later on in this chapter. Now our main aim will be to review what is known about the mechanochemical devices and processes involved in the production of traction forces which are required both for moving a cell and for making it capable of pulling the extracellular material.

These devices are mostly localized within a cell part usually defined as a leading lamella (Harris, 1994; see this paper also for further terminology). It is not always adjusted to the anterior cell pole; a 'tug-of-war' between two or several differently directed leading lamellae can take place. In any case, two main kinds of interrelated mechanochemical events seem to occur within a lamella: (1) anteriorward spreading of a 'central core' and

(2) rearward flow of more peripheral material, including subcortical actin and some membrane components.

Anteriorward spreading implies the production of a pushing force. While some authors attribute it to the anteriorly directed polymerization of an actin bundle (Alberts *et al.*, 1989), others (e.g., Bereiter-Hahn and Lüers, 1994) argue for the main role of a swelling force within the actin gel. According to the latter point of view it should be the actin depolymerization (as increasing the osmotical pressure of actin gel) which promotes the advancement of a protrusion. Probably both points of view may be combined with each other.

Another indispensable component of a cell's movement or of the exertion of a traction force is the retrograde transport of a subcortical polymerized actin together with some membrane components anchored onto it. At the rear edge of a leading lamella these actin structures are disassembed. A monomeric G-actin diffuses towards the anterior, where it is assembled again into a new cytoskeletal network (Fig. 2.9). The retrograde transport and the foreward spreading of a lamella are most likely the components of a single continuous process; there seem to be no evidences that they alter each other in time. According to the 'raking hypothesis' (Harris, 1994) a flowing actin, in its retrograde movement, involves a certain special subset of the integral proteins of a cell membrane, rather than an entire membrane as a whole. It is this 'raking' process which produces the traction forces applied either to the extracellular matrix or to the artificial substrates. The

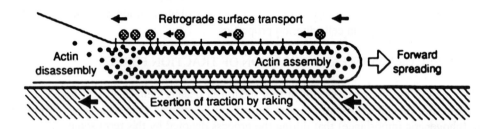

Figure 2.9. Diagramm of the relationship believed to exist between treadmilling of cortical actin and the retrograde surface transport along the plasma membrane (From Harris, 1994, with the author's permission).

dynamical processes occurring within a leading lamella show also a tendency for lateral propagation; this is especially clear in the cases of rotatory cell movements.

There are also some evidences that the leading front of the moving cells is a preferred location for exocytotic events, while mostly extensive endocytosis is related to a cell body, and in particular to the lamella transition region (Bergmann *et al.*, 1986).

Generally, our knowledge of the mechanisms involved in cell movement and in the exertion of traction forces is still very incomplete and far from being integrated into a kind of a coherent picture, although it is currently one of the 'hottest' points of cell biology and biomathematical modelling (e.g., Alt and Kaiser, 1994). What seems, however, to be already obvious is that the actin and the associated membrane flow (briefly: actin flow, or simply flow) in the crawling and traction generating cells is a perfect example of a supramolecular dynamical structure, far removed from thermodynamic equilibrium. In this respect we may use once again the image of a multilevelled hierarchy. Within such a framework, the actin flow represents by itself a new level, located considerably higher than that of the individual actin molecules. If we estimate (data taken from Harris, 1994) the average length of the region involved in the flow as 10 mkm, it will include about 1500 G-actin molecules (a diameter of an individual molecule is about 6 nm). Each single actin molecule is in a state of continuous flow for about 10 min = $6 \cdot 10^2$ s (to be compared with no more than 1 s time required for each one extreme G-actin molecule either to enter or to leave the flow). Consequently, a crawling of a fibroblast is associated with the emergence of a new dynamic level with L_{ch} of about 10 mkm and T_{ch} of about 10 min.

As shown by model experiments (Shimizu, 1984), the properties of the individual actin molecules are largely modified while being involved in a flow. This may lead to a complete modulation of an entire intracellular signal system (Forgacs, 1995), being probably a cause of drastic changes in genes' activity within the spreading fibroblasts, as compared with the freely suspended ones (review: Opas, 1994).

Meanwhile, the actin flow is not the uppermost level associated with a fibroblast motion. As we have recently mentioned, the trailing force exerted by a single cell can be spread along the extracellular matrix or an artificial substrate up to the distances of several millimeters (thus exceeding the length of a flow region by no less than 2 orders) and be maintained for hours. Such are the ranges for L_{ch} and T_{ch} characterizing an upper level's dynamic structure, including an entire cell together with a surrounding extracellular matrix. The interactions of this level events with those of an actin flow level are bidirectional: not only the flow is required for producing long-range stresses, but the latter are also important for maintaining the flow. How these long-range interactions can affect the behaviour of cell populations we shall see in the section 2.3.2.

2.2.3. DIVISION OF A CELL BODY (CYTOTOMY)

The dominant views about the mechanisms of cytotomy, mostly based upon the studies of cleaving zygotes, have evolved remarkably during the last two or so decades. The initial idea was that the main motive forces are located within a furrow region and are exerted exclusively by a sphinctor-like contraction of a ring of actin microfilaments, which are arranged, under the influence of a mitotic spindle, just within the furrow plane. What has definitely remained from this viewpoint was the role of the spindle elements, namely astral microtubules, in arranging actin subunits and in determining thus the furrow plane. On the other hand, more recent investigation made it obvious that the activities associated with a cytotomy were in no way restricted by the furrow region itself, but involved also a subcortical area of an entire egg.

Most of authors (White and Borisy, 1983; Shroeder, 1990; White, 1990) distinguish from each other two successive steps of a cytotomy. The first, which is independent of a mitotic spindle, implies an overall activation of the subcortical contractile elements (F-actin), leading thus to the development of a substantial tension. Its indispensability for further furrowing has been demonstrated by arresting cleavage in zygotes relaxed as a result of incubation into a hypertonical medium (Harris, 1994a). The second phase is associated with a redistribution of the subcortical tensions in such a way that a polar–

equatorial tensile gradient emerges, with its highest point located just within the furrow (i.e., equatorial) region. The polar regions are assumed instead to be either actively expanded, or at least relaxed. Such a gradient is believed to drive the flow of F-actin subunits towards the furrow region. In this region they are accumulated and take a parallel alignment. As a result the tension in the furrow plane should exceed that in the perpendicular plane by an order or more (Harris, 1990).

Although such a scheme cannot be considered as being fully approved, its similarities with a much better tested dynamical scheme of a single cell's movement deserve to be mentioned. In both cases tensions on the surface are indispensable, and both imply an overall flow of a subcortical actin. Within such a framework the polar regions of a cleaving zygote can be considered as being homologous to the leading edge of a crawling cell, while the equatorial regions to its trailing end, or to the posterior base of the leading lamella (Fig. 2.10). This view is reinforced by the observations of an exocytotic activity in the polar regions of a zygote and the endocytosis in the furrow region (Kappel *et al.*, cited from White, 1990; Mescheryakov, 1991).

Actin structures within the furrow region also cannot be considered as a simple muscular sphincter. Firstly, its purely 'contractile' nature is a matter of doubt in the cases of unilateral furrows, such as those in Coelenterates and insects eggs; here the contractile band should produce more 'pushing', rather than contraction (Schroeder, 1990). Also, a very rapid (several minutes time) assembly and disassembly of a contractile ring is hardly compatible with its work as a coherent sphincter. It is probably just F-actin depolymerization, coupled with membrane endocytosis, rather than F-actin contraction,

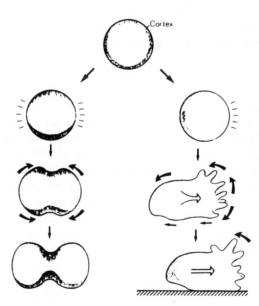

Figure 2.10. Diagram illustrating a dynamic similarity between a cleaving cell (left column) and a migrating cell (right column). In the cleaving cell the lower cortical tension is in two polar regions, while in a migrating cell it is at the leading edge. In the both cases a cortical material is flowing towards the highest tensions sites (from White, 1990, with the publisher's permission).

which plays a major role in many cases of a cytotomy. Mescheryakov went even further by neglecting, according to his electron microscopy observations, the very existence of an integrated contractile ring in zygotes of a pond snail, *Lymnaea stagnalis*. He claimed that neither conventional myosin, nor microfilaments are concentrated in a *Lymnaea* cleavage furrow. Instead (Mescheryakov, 1983, 1991) he observed a coherent submembrane actin layer, perforated at many sites with small pores lacking microfilaments. In the isolated preparations of the cortical layers this layer looked as consisting of the peculiar domains of the mutually parallel actin microfilaments, each of them comprising filaments of identical length and polarity. While observed in a radial view, these domains have been arranged as twisted helicoids, laying one upon other and similar to cholesteric liquid crystals (Figs. 2.11, 2.12). In most cases they consisted of 4 stores of the helicoids, spaced by 15–30 nm. Each next store microfilaments were usually twisted relative to a neighbouring one by a considerable screwing angle. These helicoids could be seen in all the regions (both polar and equatorial), but were mostly concentrated in the equatorial regions. The author suggests that these remarkable structures are somehow involved in generating dissymmetrical (either clockwise or counterclockwise) shearing shifts of the extraequatorial regions of a zygote membrane, which he observed at the anaphase stages of the cleavage divisions. A seemingly simple process of an early egg's cleavage contains still so many enigmas! But the main lesson is obvious and similar to that derived from a cell crawling analysis: in a cleaving egg we are dealing again with a powerful cortical-subcortical dynamic structure which includes the microtubules assembly, actin flow and membrane turnover. This structure extends over an entire egg diameter (that is, up to millimeters) and has a life time no less than several dozens of minutes.

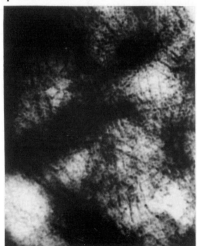 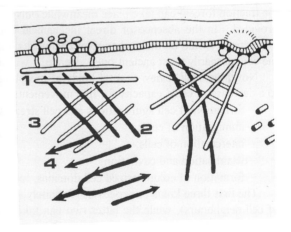

Figure 2.11. The multistored microfilament domains in the membrane ghosts of a pond snail (*Lymnaea stagnalis*) zygote. Decoration by meromyosin (From Mescheryakov, 1991, with the author's permission).

Figure 2.12. Organization of the microfilaments sub-membrane layer in the pond snail zygote. Numbers designate microfilament domains, their polarity being shown by thick arrows in two lower domains. Domains NN 1-4 form a stepwise twisted pillar - a helicoid.

2.3. A morphodynamics in cell collectives

Now we pass to the morphogenetic events which take place in large cell collectives. As previously, we will be interested most of all in exploring their multilevelled dynamic structure, interlevels feedback, and self-organizational capacities of these events. In our account we shall distinguish two extremal categories of cell collectives: freely moving mesenchyme-like cells on the one hand and tighly packed epithelia with a very much reduced capacity for exchanging cell neighbours on the other. Inbetween we should insert the recently explored, and probably most widely spread, cases of an extensive cell repacking within rather dense cell masses.   As to the epithelia, their main morphogenetically important properties are as follows:

(1) the abovementioned capacity to maintain the polarized (apico-basal) ionic transport; owing to this, an excessive turgor pressure is produced within the subepithelial cavities, this very pressure stretching the covering layers. It is this moderate stretching, as a rule averaged over considerable distances, which should be regarded as one of the necessary initial conditions for morphogenesis.

(2) extensive, fast, and cooperative contact cell interactions; among them one of the most important is a so called contact cell polarization, to be discussed later on in more details.

For the mesenchyme-like cells a morphomechanical situation is very much different. The direct contact interactions via adjacent cell membranes are to a great extent replaced by more distant interactions either mediated by an extracellular matrix or spread within a matrix-free space. Considerable mechanical effects (the abovementioned propagation of the traction forces) can take place meanwhile only if a proper matrix with some elastisity is provided. In the absence of direct mechanical connections the only effective kind of interactions would be the exchange of chemical signals. Such a way seems to be amongst the evolutionarily most ancient ones. Hence, it will serve us as a starting point.

Now we shall review several kinds of morphogenetic processes, arranged according to a decrease of cells' capacity for free movements:

– formation of cell aggregations and cell streams;

– immigration of cells;

– intercalation of cells;

–delamination and cavitation;

– formation of columnarized cell domains, 'cell fans' and the epithelial folding;

The first three kinds of processes definitely imply free cell movements (the exchange of cell neighbours), while the latter two can take place at the fixed positions of adjacent cells (although some amount of cell repacking may also take place).

2.3.1. CELL AGGREGATIONS AND CELL STREAMS IN SLIME MOULDS

Morphogenesis of slime moulds (Acrasiaceae, *Dictyostelium discoideum*) is one of the most thoroughly studied developmental processes. Let us recall briefly its main characteristics according to Devreotes *et al.*, 1983; Durston, 1973; Gerisch *et al.*, 1977; Robertson and Grutsch, 1981; Vasieva, 1991.

As is well known, single cells (myxamoebae) of *D. discoideum,* which on a nutritive substrate are living separately from each other and proliferate for an indefinitely long time, start to aggregate after being seeded onto a hunger substrate. It is the process of aggregation which will be of most interest for us.

Within first hours after seeding onto a hunger substrate an unstable reversible aggregation into small cell clusters takes place; later on, in 8–9 hrs, more organized and irreversible collective aggregation starts. All the kinds of aggregation are based upon chemotactical cell movements, directed upwards along the gradient of an attractant, which is usually cyclic adenosinemonophosphate, 3'5'cAMP. If the gradient is steep enough (no less than $2 \cdot 10^{-8}$ M/mm), cAMP-dependent Ca^{2+}-channels become activated at the leading edge of a cell, causing a local rise of intercellular $[Ca^{2+}]$. Along with that, the concentration of another cyclic nucleotide, cGMP is rised within the cytoplasm of the activated cells. Both events switch on the intercellular motile machinery.

At the beginning of aggregation cAMP is secreted by all of the cells spontaneously, permanently, and in small concentrations. Later on a more effective relay regime of cAMP secretion is established. Now, the cells secrete cAMP by periodic impulses which follow one after another in several minutes (from 10 minutes at the beginning of aggregation to 2 minutes before its completion). Within the same time period the concentration of cells secreting cAMP rises in more than 4 orders, each one cell producing and secreting cAMP only in responding to the external cAMP impulse. On the other hand, a cell perceiving the external cAMP becomes immediately insensitive to other cAMP impulses; such a refractory time lasts up to 8–10 min. During these 'deaf' periods the cells are moving directly towards a source of an external cAMP at a rate of about 20 mkm/min. It is due to the relay secretion that the cells can collect themselves into a single cluster from a territory of no less than 1 cm², which exceeds by two orders the distance of a diffusional cAMP spreading from an individual secreting cell (the latter distance is about 100 mkm).

In 8–11 hrs from the start of starvation a cell population is structured: one can see the oriented cell chains directed towards aggregation centers (Fig. 2.13 A). At the early stages of aggregation the chains can be easily dissolved or restructurized; some unstable microvortices appear, each of them involving a single chain or a few chains of cells (Fig. 2.13 B,C). Later on the cell streams are stabilized, small branches disappear, and within the streams of larger diameters some compact beads can be seen, moving towards the aggregation center (Fig. 2.13 D). In some cases the occasionally detached beads become the new centers of attraction. Cell streams take, as a rule, the shape of turbine blades. An aggregation center is either a dense rounded cell accumulation or a rotating torus – Fig. 2.13 E,F (Vasieva, 1991).

At the end of aggregation the cells pass from an induced towards an autooscillating cAMP secretion regime. That means that each cell starts to produce cAMP autonomously in a periodic regime. Later on a period of constant cAMP secretion is initiated. At this time the cells become connected with tight junctions. In such a way a common pseudoplasmodium is formed out of several thousands of aggregated cells. Long before it has been transformed into a fruit body, its cells are differentiated into two types, which are called the prestalk and prespore ones.

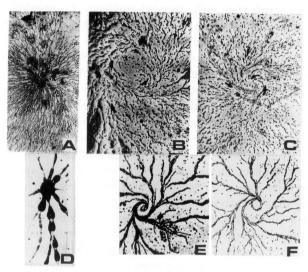

Figure 2.13. Aggregation patterns in slime moulds. A: radial cell chains at the early stage of aggregation; B, C: unstable vortices at the advanced stages; D: "beads" within the aggregation streams; E, F: "turbine blade" structures created during the latest aggregation period (from Vasieva, 1991, with the author's permission).

How can we comment upon the aggregation process from the general point of view? First of all, it should be qualified as a 'pure' self-organizing process, insofar as there exist no prepatterns of any kind for the cell clusters arised. The cell clusters, streams, and other structures created by the myxamoebae can be naturally considered as the singularities of a common non-linear field, their appearance being a result of the loss of stability of an initial homogeneous state of the field (Belintzev, 1991). Meanwhile, before discussing a suitable mathematical model of the process let us explore, what its dynamical levels may look like. Two such levels at least can be definitely distinguished. The lowest one corresponds to single cells activity. Its T_{ch} is about 2 –10 min = 10^2 s (this is the period between two successive cAMP impulses; same is the refractory time), and L_{ch} is about 100 mkm = 10^{-4} m (such is the approximate diameter of a diffusion field exerted by a single cell). On the other hand, the times required for the abovedescribed evolution of the aggregational regimes are much greater, namely of the order of several hours. As a result, extensively enlarged aggregational territories appear, up to 1 cm diameter. In such a way an upper level is created, possessing T_{ch} of the orders of 10^4 s and L_{ch} of the orders of 10^{-2} m. There are the events associated just with this upper level (namely, the formation of cell chains) which substantially increase the overall aggregation rate.

Both levels are based in Acrasiacea upon the distant chemical interactions. This is but natural because at the beginning of cell aggregation neither cell-cell contacts, nor a common elastic substrate, required for the mechanical interactions are presented. It is of interest, however, that as soon as the intercellular contacts are established and the stream structures created such interactions immediately come into play. Thus the formation of the abovementioned beads is obviously the result of the interfacial tension between a cell stream and the surrounding medium. In any case, the diameter of the beads have nothing in common with the length of a diffusional cAMP wave (Vasieva, 1991). A cylindrical shape of a pseudoplasmodium can be interpreted within the framework of a stress theory of growth (see p. 93, equation *(17)* and the corresponding comments). A typical shape of a fruit body can be also regarded as a result of the balance of a chemotactical attraction of the cells towards the tip and of the tension on the surface of a fruit body (Belintzev, 1991).

However, a general model of the events taking place during the aggregation in slime moulds, which was suggested firstly by Cohen (1977) and elaborated further by Belintzev (1991), is based purely on chemokinetical notions. Its main ideas, formulated in the terms of the abovedescribed two levels, are as follows.

The lower level processes, that is, the periodic secretion of cAMP by the individual cells, are assumed to be based upon the following mutually competing events: (1) cAMP synthesis, going with a saturation and reaching thus a stationary level; and (2) a spontaneous splitting of cAMP, revealing a linear dynamics. In a rather general case, the resulting concentrational curve will be S-shaped (Fig. 2.14), thus describing a spontaneous periodicity of a certain rate. The corresponding equation will look like

$$\frac{dc}{dt} = f(c,h) - qc \qquad (18a)$$

where c is the concentration, q is the parameter of the cAMP splitting rate, while h is up to now non-identified parameter.

For imitating the upper level events we should introduce the second equation, describing in a rather simple (linear) way the h parameter:

$$\frac{dh}{dt} = nSc - ph \qquad (18b)$$

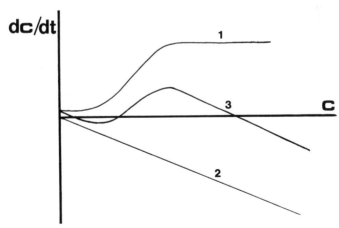

Figure 2.14. Rate of cAMP autocatalysis (dC/dt) as a function of its concentration (C). 1: saturation curve; 2: cAMP splitting; 3: a resulted S-shaped curve obtained by a summation of 1 and 2.

Before discussing the meaning of the other parameters introduced and the entire equation *(18a)*, let us take into consideration that the equation *(18b)*, as being dependent upon the value of the S parameter, can modulate the fundamental lower level non-linear dynamics in standard ways, already described in Part 1 . Namely:

(a) at the highest S values we get rather stable and stationary, but low level cAMP production (Fig. 2.15 A);

(b) by diminishing S we come towards an unstable point, where a small perturbation is enough for producing a high cAMP pulse (Fig. 2.15 B); that corresponds to a relay regime;

(c) by continuing to rotate the linear isocline (decreasing S value) we come to the autooscillations regime (Fig. 2.15 C),

(d) and, finally, we come to the regime of a spontaneous high rate cAMP secretion (Fig. 2.15 D).

Now, what might be the biological meaning of the h and S parameters? h can be regarded as an inhibitor of cAMP synthesis, while S may be considered as a stored reserve of the inhibitor's precursor. Within such a framework, the reduction of S value can be interpreted as an exhaustion of the precursor during continuous cultivation onto a starved substrate. Thus, a slow dynamics of the changes in cAMP secretion regimes looks to be quite natural. Note, meanwhile, that, contrary to higher organisms, the main controlling parameter here comes from the external environment.

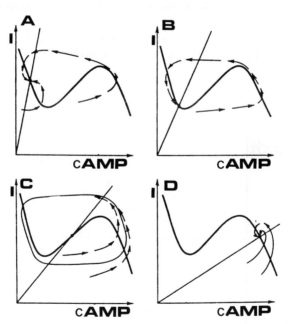

Figure 2.15. Phase portraits for the dynamic schemes of cAMP synthesis in slime moulds under four different values of a control parameter S (eq 2.10b). Horizontal axis: cAMP concentration, vertical axis: concentration of an inhibitor, I. A: homeostatic equilibrium under low cAMP level; B: "waiting regime" (relay activity); C: a regime of an autonomous rhytmical secretion; D: homeostatic equilibrium under the relatively high cAMP level (from Belintzev, 1991, with the author's permission).

We may conclude that a great deal of developmental dynamics in slime moulds definitely fits quite simple, canonical models of self-organization, up to the direct imitation of the Beloussov–Zhabotinsky reaction structures. Is there however so much in common between this peculiar group of organisms and the real Metazoans? As we shall see soon, the latter only partially (and with much greater difficulties) can be described by the canonical models of a self-organization theory. Some very important evolutionary novelties should appear on the evolutionary way between Myxomicetes and Metazoans morphogenesis. It is of a interest to realize, meanwhile, that the first multicellular organisms created by the Nature behaved as diligent students of a self-organization theory and only their evolutionary descendants permitted themselves to deviate from its canonical ways.

2.3.2. AGGREGATIONS AND STREAMS OF MESENCHYMAL CELLS IN METAZOANS

In Chapter 1 we have already talked about a 'spontaneous' structuring in the fibroblast colonies, and have mentioned, that they could form regular aggregations only in the absence of a collagenase (that is, in the presence of a collagen). That indicates a crucial role of an extracellular matrix in cell aggregation. In a most straightforward manner such a role has been demonstrated by Harris and coworkers (Harris et al., 1980, 1984; Stopak and Harris, 1982;) with the use of artificial elastic substrates, (collagen gels or silicons), capable of bearing and transmitting to considerable distances the above mentioned traction tensions, including even those produced by single cells. The main result of these experiments was a splitting of an initially homogeneous fibroblast culture into a number of compact, radially wrinkled cell clusters, connected by the files of extensively stretched cells (Fig. 2.16 A-C). Here we are dealing for the first time with the self-maintaining and self-reinforcing structures associated with the coupled compression–extension stresses. A lack of a similar structuring onto a non-deformable substrate strongly argues for the crucial role of the mechanical factor, rather than, say, chemotaxis, in creating *de novo* a morphological pattern.

The entire process was given a refined mathematical description (Oster *et al..*, 1984) from which we will extract only the leading ideas (let us denote this concept as OMH model, by the first letters of the author's names).

Three main factors which determine cell behaviour (besides cell proliferation and cell death) are implied in the OMH model. The first of them is a short-range cell adhesion, that is, a cell's capacity to establish firm contacts over the distances not exceeding their fillopodia length. Such a factor would obviously promote cell condensation. The second factor is the so called haptotaxis, or a cell's tendency to move up along the adhesion gradient. This can be considered as an almost inevitable consequence of the adhesion itself, since if in the direction A a cell could establish more contacts per time unit than in a direction B, it will obviously shift itself towards A.

The only long-range factor to be introduced is the stretching of a substrate, caused by the cells themselves, and bringing into being a passive drift (convection) of the same cells along the substrate.

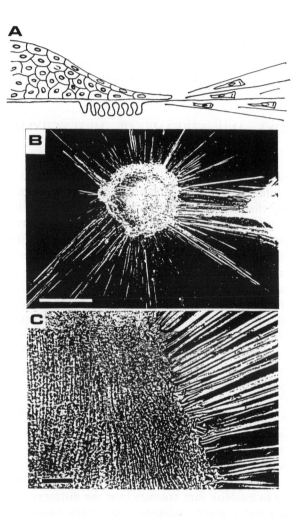

Figure 2.16. Fibroblast onto an elastic (silicone rubber) substrate. A: a diagrammatic side view of the margin of an explant whose cells are spreading outward on a silicone rubber substratum. The traction forces exerted by the outgrowing cells compress the rubber sheet beneath the explant and stretch it into long radial wrinkles in the surrounding area. B: a dark-field illumination view of a chick heart explant that have been spread onto a substratum for 48 h. The bright radiating lines are stress wrinkles. C: higher magnification, view of the marginal outgrowth zone of the same explant. The compression folds beneath the explant are seen on the left an the radial stress wrinkles around it are on the right. The bar is 100 mkm long. (From Harris et al., 1980, with the author's permission).

Let us suggest now that within a certain small area the cell concentration has been occasionally increased. As a result a short-range adhesion gradient would appear, providing, owing to a haptotaxis, further local increase in cell concentration. The cells which move towards the aggregation center stretch the adjacent area of a substrate (extracellular matrix) together with the attached cells. At the opposite border of the stretching zone a new short-range gradient of the cell density will appear, oriented oppositely to the first one. Here again haptotaxis will come into action, producing the second cell condensation, etc..

The main bifurcation parameter of the OMH model is the rate at which the cells stretch a substrate. Only after this rate exceeds a certain threshold a uniform cell mass will be split into areas of an increased and decreased density.

Let us now compare OMH model with those of a Gierer–Meinhardt type (GM models). Both are similar to each other in introducing the short- and long-range factors and ascribing to the former the activating, whilst to the latter the inhibitory functions. On the other hand, only within the OMH model's framework both kinds of factors look like being the natural consequences of an entire situation, while in GM models each of them is introduced separately, as if 'by hands'. Also, the OMH model uses tensor equations, which give a larger class of singularities rather than the vector equations employed in the GM model. And what is for us of an utmost importance, the OMH model widely employs the notions of essentially mechanical feedbacks, related to two different levels: the lowest corresponding to the single cell scale and the uppermost belonging to a macromorphological scale, its L_{ch} being in the range of millimeters. As concerning the 'natural' morphogenetic processes, the OMH model can be used for interpreting the formation of vast mesenchymal patterns associated with the formation of feathers or hairs in vertebrate embryos (Oster et al., 1984).

2.3.3. IMMIGRATION AND INTERCALATION OF CELLS: TOOLS FOR REMODELLING THE SURFACE/VOLUME OR THE LENGTH/SURFACE RATIOS

These two important morphogenetic processes take place in more or less dense masses of cells which are nevertheless capable of exchanging their neighbours. Such an exchange ('repacking' of cells) is just what is in common between both kinds of processes, which lead nevertheless to almost opposite global morphological results. Immigration, which means the active movement of cells from the surface of a cell mass towards its interior (Figs. 2.17, 2.18 C) is inevitably associated with a *decrease* of the surface/volume (S/V) ratio of a sample, while intercalation leads, instead, to an *increase* of either S/V ratio, or/and the linear/surface (L/S) ratios. An S/V ratio increase (that is, a uniform flattening of a sample) is caused by so called radial cell intercalation (Fig. 2.18A), while an L/S ratio increase means a sample's extension along a certain direction owed to a so called planar, or convergent cell intercalation (Fig. 2.18B).

As it was emphazised by Keller (1987), cell intercalation is one of the most universal morphogenetic tools for achieving either isotropic expansion or, more often, directed elongation of quite various rudiments and body parts, beginning from a fresh water hydra (Bode and Bode, 1984) and marine hydroids (see later), including insect imaginal discs (Fristrom, 1976), a gut rudiment in Echinodermata embryos (Ettensohn, 1985;

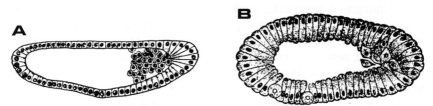

Figure 2.17. Immigration of the endodermal cells in Coelenterates embryos is going on from the narrow (rightwards oriented) poles of the planulae larvae. A: *Tiara pileata*. B: *Mitrocoma annae* (from Zachvatkin, 1949).

Hardin and Cheng, 1986) and reaching its full expression in vertebrate embryos, where it is qualified as 'the main engine of gastrulation' (Keller, 1987) and of the axial organs formation. In the latter case we are dealing with the convergent (latero-medial) intercalation, the intercalating cells moving in the convergent directions along a sheet plane towards the dorsal embryo midline (Fig. 2.18B). Inspite of a great amount of randomness in the individual cells' movements (as already mentioned in Part 1) the macroresult is always quite uniform and stable. It is, in the first place, just the convergent cell intercalation which moulds an entire vertebrate's body, by elongating it in the direction perpendicular to the intercalation movements of the individual cells.

What we would like to argue now, looks at first glance paradoxical: it is that both immigration and intercalation, inspite of their opposite macromorphological results, are practically identical in their local aspect (that is, from the 'point of view' of single cells),

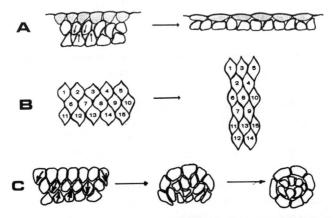

Figure 2.18. A comparison of cell intercalation and cell immigration in amphibian's embryonic tissues. A: anisotropic expansion of an element of embryonic surface (blastocoel roof) because of the radial intercalation of the deep cells (arrows). The superficial ectodermal cells (hatched) are passively stretched. B: convergent intercalation of the suprablastoporal zone cells, leading to anisotropic extension. C: sphericalization of an isolated piece of a tissue by means of the same local cell rearrangements as in A, but implying other boundary conditions (namely, existence of free edges). As a result, the intercalation is transformed into immigration. A, B from Keller, 1987, modified. C is original.

insofar as both are associated with the cells' tendency to increase the areas of their mutual contacts and to diminish, correspondingly, their contacts with a cell-free environment. As to the oppositeness of their end-results, it is caused by the presence of the global mechanical constraints to a free rolling (folding) of a cell sheet in the case of intercalation and by the lack of such constraints in the case of immigration. This means also that the tangential stretching of a cell sheet would promote intercalation, while in the lack of stretching (low tension on the surface) a piece of embryonic tissue will be biassed towards immigration.

While starting to substantiate this, up to now, speculative claim, we may address ourselves firstly to the early embryos of hydroid polypes, pointing out that the immigration areas coincide here with the sites of a greatest curvature (Fig. 2.17): those are, according to the Laplace law, the areas of lowest tension. On the other hand, the intercalation-like processes proceed in these animals in the mostly stretched tissues, and are positively correlated with the amount of stretch (see Chapter 3 for more details). Meanwhile, in a mostly straightforward manner a choice between the intercalation (S/V ratio increasing) and immigration (S/V ratio decreasing) pathways can be traced in the blastocoel roof tissue of a blastula or early gastrula amphibian's embryo. Normally, so called deep ectodermal cells of this region undergo a radial intercalation, thus reducing a number of cell layers and, at the same time, extending an entire sheet(Fig. 2.18 A). This is the case, however, only if the region considered is prevented from a free rolling or from any possibility to diminish the existing tension. Being, instead, permitted to roll, a piece of a tissue becomes, within a couple of hours, transformed into a solid ball with an excentrically located cavity (a small model of an entire blastula!), hence considerably reducing its S/V ratio (Figs. 2.18 C, 2.19, A-C). The same trend, also associated with a considerable cell immigration, can be traced in the entire blastula stage embryos with relaxed tangential tensions (see 3.3.1, pp. 179-182).

On the other hand, an artificial unidirectional stretching of the same tissue piece evokes in it convergent intercalation movements to which these tissues are normally

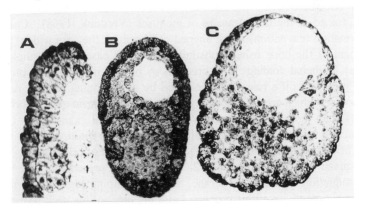

Figure 2.19 A-C. Formation of a blastula-like explant from an isolated piece of animal ectoderm in *X.laevis* embryos at the early gastrula stage. A: 10 min, B: 5 h, C: 18 h after isolation. Note *de novo* formation and the subsequent enlargement of a blastocoel-like cavity in B and C.

incapable (Beloussov and Luchinskaia, 1995; see 2.5.1.for more details). Also, the convergent intercalation movements in the isolated pieces of suprablastoporal regions can be reoriented perpendicularly to their normal (latero-medial) direction if stretching a piece in this very direction (that is, transversely rather than longitudinally) – Beloussov *et al.*, 1988. It is worth mentioning that in both series of experiments the intercalation occurred as a *post*-stretching reaction, that is, already after a piece has been fixed in a stretched state. That means that the stretching acts here both as an initial (and boundary) condition and as a governing parameter.

In any case, the stretching or other constraints of free rolling definitely belong to a macroscopic (supracellular) level. This follows not only from the dimensions of the stretched areas, but also from the duration of a required stretching, which in no case is less than 30–40 min (to compare with the few minutes required for each single cell to change its neighbours). We may conclude, therefore, that intercalation, in order to be initiated, requires regulation from an upper level, while immigration, which employs practically the same local cell activities, is, instead, promoted by a weakening of the upper level's constraints. This is closely correlated with the fact that the intercalation is directed towards a decrease of symmetry order of a macroscopic system, while immigration, on the contrary, is directed towards its increase.

2.3.4. DELAMINATION AND CAVITATION: PATTERN FORMATION BASED UPON DIFFERENTIATION OF THE STRESSED NETWORKS

In embryology the term 'delamination' is usually reserved for describing the mode of the separation of the germ layers, associated with a straightening (alignment) of their borders without more or less extensive cell movements and repacking. Actually, the processes of alignment of cell walls are not restricted by the segregation of the germ layers only, but are taking place in a lot of other cases and taxonomic groups. Their roles in delimiting from each other so called 'cell polyclones' in insect development have been extensively discussed elsewhere (Crick and Lawrence, 1975; Karlsson, 1984), and similar events have also been traced in the late cleavage of molluscs (Verdonk, 1968). Coupled often with cavity formation (cavitation) along aligned cell borders, they are widely presented in the development of the bone fishes, substituting in many cases (for example, in neural tube formation) epithelial folding. The preciseness of the delamination processes is really amazing: it looks as if a ruler had been used by a developing sample. And, nevertheless, these processes which are so important and remarkable remain quite poorly studied.

Let us start their description from tracing a classical example of the delamination, taking place in solid embryos of a hydroid polyp, *Clava multicornis*. A little time before the start of delamination some randomly situated groups of cells appear, bearing peculiar 'necks', spread over each other (Fig. 2.20 A, B). The presence of the "necks" indicates an overproduction of a cell membrane, and the grouping together of neck-bearing cells points to a cooperativity of this process. Later on, most of the neck-bearing (oblique) cells become located on the edges of an already delaminated area (Fig. 2.20 C, D). An entire delamination process looks like a progressive enlargement and fusion together of the already delaminated pieces, up to their closure into a complete circle (Fig. 2.20 C-E). That makes it probable, that the delamination is going from one already 'aligned' cell to

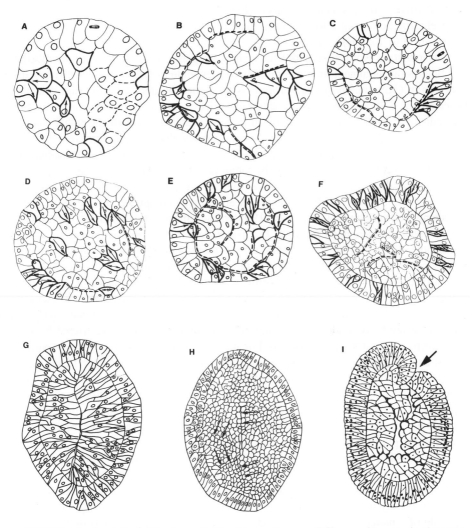

Figure 2.20. Delamination in hydroid polypes. A-G: subsequent stages of *Clava multicornis* development, from morula up to a planula stage. Outlined are the 'neck-bearing' cells, arrows indicate the directions of the necks formation. Dotted are more or less aligned segments of cell borders, sometimes abortive and later on disappearing. H: first vague outlines of a *Tubularia* embryo petals are associated with alignment of tangential and radial (arrows) rows of cell walls. I: *Dynamena pumila* embryo just after accomplishment of a delamination. Arrow indicates the initiation point of a delamination relay, coinciding with a future posterior pole of a larva. Note the differences in the curvatures of the opposed sides of a larva simulating its dorso-ventral patterning. A-G: original, H: courtesy of J. Krauss.

another as a kind of a relay process, the involved cells often passing via an oblique shape. Somewhat later we shall present more direct evidences supporting this suggestion, using amphibian embryos as an example. What we would like to notice now is a lot of topographical 'errors', while delamination is on the way (Fig. 2.20 B, D, F). The same is true for the subsequent formation of a gastral cavity , or cavitation, which also starts from some incorrect (in their orientation) 'aligning attempts' within an endoderm. Several such abortive attempts can be traced (Fig. 2.20 F) before one of them, namely that precisely oriented along a longitudinal embryonic axis which is just forming, will become successful (Fig. 2.20 G).

Delamination is not just a tool for separating germ layers. One of its morphogenetic functions is the alignment of some, more or less extended segments of embryonic surfaces. Whereas in *Clava* embryos the polygons with aligned borders arised during delamination process (Fig. 2.20 F,G) are only transitorial and leave no traces in further development, in the radialized *Tubularia* embryos similar polygons associated with the radial aligned cell borders (Fig. 2.20 H, arrows) create a germ of an initially vague and non-precise but then self-focusing petal pattern of an actinula larva. Another function of the delamination in hydroid polypes is to irreversibly fix the longitudinal axis and, in some cases at least, also the antero-posterior polarity of a larva. Whereas prior to delamination a presumptive longitudinal axis can be easily reoriented by an artificial compression (Beloussov, 1973) or stretching (Krauss and Cherdantzev, 1995) of an embryo, this becomes impossible after the delamination have been resumed. In *Dynamena pumila* embryos the site of a posterior larval pole (Fig. 2.20 H, arrow) was shown to be causally correlated with the site of initiation of the delamination wave (Krauss and Cherdantzev, in press). Interestingly, during the delamination two opposite sides of *Dynamena* larva acquire different curvatures thus imitating a kind of a dorso-ventrality. These curvature differences well may be the direct results of a passage of an alignment wave along the surface of an incompressible body; under these conditions the flattening may be restricted by the area adjacent to a site of a wave initiation. For most Coelenterates such a 'dorso-ventrality' is almost useless (with possible exception of some crawling larvae) but may be regarded as a kind of an 'evolutionary preview'. It indicates also an inherent capacity of a delamination process to create coherent shapes of a reduced symmetry order without any additional instructions, either from inside or outside. As viewed in such a manner, the delamination should be really attributed to the self-organizing processes, generating coherent macroscopic patterns out of a largely imprecise backround.

It is possible to suggest therefore that the morphogenetical results of delamination very much depend upon the number of its initiation points and the rates of the delamination wave(s). Most probably, the number of points, as well as the waves rates, are not precisely determined from the very beginning. Let us take a very much simplified example of a planar incompressible body, with its initial shape being close to circular and suggest (as a mostly general assumption) that delamination is spread from its initiation point(s) with unequal rates towards both sides. Then, in the case of one initiation point only, we shall obtain (in accordance with the above presented expeimental data) a polarised and 'dorso-ventralized' shape with the symmetry order $1 \cdot m$ (Fig. 2.21 A_1 - A_3). Having two initiation points situated opposite each other we obtain a shape which is closer to a rotationally symmetric one (symmetry order $\infty \cdot m$), Fig. 2.21 B_1 - B_3. At last, in the case of n initiation points, the resulted symmetry will be an n-order radial one (Fig. 2.21 C_1 - C_2). It is obvious, on the other hand, that the greater be the number of initiation points, the less will be the deformation of a resting (still non-delaminated) part of an embryo, resulting from its delamination-caused alignment. Would it be possible to derive a number of initiation points as a function of an allowed deformation? This question still remains open.

From a more local (cellularly oriented) point of view, we can easily see that the delamination belongs to a set of processes which may be defined as a 'differentiation of tensions', that is a segregation of a net consisting initially of more or less equally and isotropically tensed cell walls into two extremes characterized by high and low tensions. Later on (2.5.1) we shall present some suggestions about how that may take place. Now we would like only to present some examples of this widely spread process as it takes place in amphibians embryonic tissues. They relate either to the normally developing

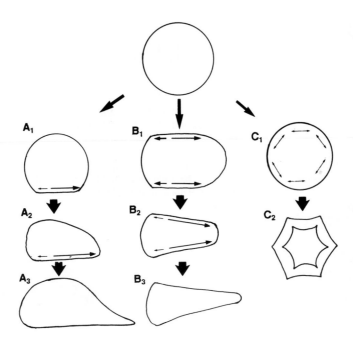

Figure 2.21. A possible involvement of a delamination into the embryonic patterning. Left column: unilateral delamination. Middle column: bilateral delamination. Right column: delamination initiated in 6 different points.

embryos (Fig. 2.22), or to the artificial combinations of two different tissue pieces fused together (Fig. 2.23 A,B). In the latter case, a relay-like fashion of the alignment spreading is quite obvious. That makes delamination similar to another related process of cells' transformation, a so called contact cell polarization, to be treated in the next section.

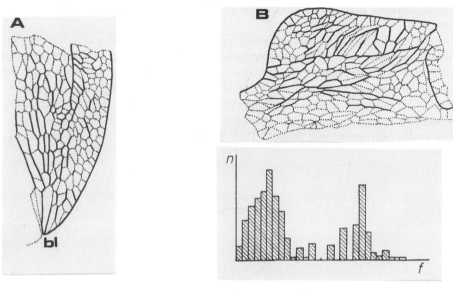

Figure 2.22. Segregation of a cell walls' net into slightly and highly tensed fractions.   A: a suprablastoporal zone of *Rana temporaria* early gastrula embryo. bl - rudiment of a blastopore. B: a left part of an early neurula dorsal zone, transversal section. Outlined are the rows of smoothed (highly tensed) cell walls. C: a markedly bimodal histogramm of the cell walls tensions in the region shown in B. Relative tension values have been estimated according to equation (2.7). The left peak indicates slightly tensed cell walls (dashed in B) while the right peak highly tensed and smoothed cell walls outlined in B (From Beloussov and Lakirev, 1988).

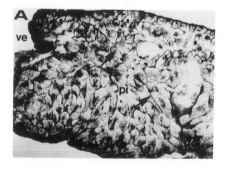
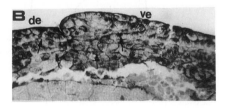

Figure 2.23. Examples of a transmission of an alignement relay from one fused piece of embryonic tissue to another. A: a fusion of a ventral ectodermal tissue (*ve*) with that of a primary inductor (*pi*). Aligned cell borders are passed continuously from *pi* to *ve* (arrows). B: a tangential fusion of a dorsal ectoderm (*de*) with the ventral ectoderm (*ve*). At least two lines of the aligned cell borders are passed from *de* to *ve* (arrows). From Beloussov (1978).

2.3.5.FIRST STEPS IN SELF-ORGANIZATION OF THE EMBRYONIC EPITHELIA: FORMATION OF THE COLUMNARIZED CELL DOMAINS AND 'CELL FANS'

Practically all the active shape changes (that is foldings, accompanied or not by a subsequent closure of a fold into a tube) of embryonic epithelia take place within continuous groups of extensively columnarized cells, that is, those elongated perpendicularly to a layer's surface: the ratio of their length (height) to the width, namely, L/W ratio of these cells definitely exceeds 2. We will denote these cell groups as the domains of the columnarized cells, or briefly 'domains'. Sometimes (as in the case of a neural plate of the vertebrates' embryos) a single domain is very much extended and is used during subsequent development for creating a number of separate folds and vesicles. In most cases, however, (sensory placodes of the vertebrates, various endodermal rudiments, etc..) each domain is later on transformed into a single fold. In all the cases the domains are formed from an epithelial layer with its cells having a uniform and rather low L/W ratio (as a rule, slightly exceeding 1); simultaneously with an extensive increase of L/W ratio of the cells entering the domain, other cells of the layer, situated to the side of it become flattened (their $L/W < 1$). Let us call this process a segregation of the domains. The segregation should be qualified as the first visible and probably the most fundamental and universal step of epithelial morphogenesis. Therefore, its proper interpretation is of a primary importance for the students of morphogenesis.

Looking more attentively at the domains immediately before their folding, one can easily note (as was firstly done long ago by Gurwitsch, 1914), that already before the start of a domain's folding, its cells axes, instead of being oriented parallel to each other and precisely perpendicular to a layer's surface, are considerably inclined, in such a way that if drawing a curve which will be perpendicular to their prolongations, it will coincide with a future configuration of a fold, to be reached, during normal development, in an hour or so; consequently, the prolongations of the cell's axes are somewhat converged towards each other (Fig. 2.24 A, B). Gurwitsch denoted this cell orientation as a 'prognostic' one, having in mind that it 'predicts' a shape of the rudiment to be yet achieved. Such a prognostication of a cell's orientation can be observed during the entire folding process: the curvature of the drawn line is always greater than that actually reached by a folding part at a given moment of time. We will define such cell groups with converged axes as 'cell fans'. Generally, each single cell domain is passed, immediately before as well as during its folding, via a more or less prolonged and pronounced cell fan stage. This is a truly universal event which can be traced, by our personal experience, in all the epithelial layers destined soon to change actively their shapes (Fig. 2.24 B-H; see also Fig. 3.7 D).

A universality of a cell fans' stage raises many questions and should largely affect our views about epithelial morphogenesis. However, we shall postpone the main discussion of these topics until to the next section, concentrating our attention now only on the problem of the segregation of the cell domains. Meanwhile, many of the conclusions to which we will now come, can be applied to the 'fan problem' as well.

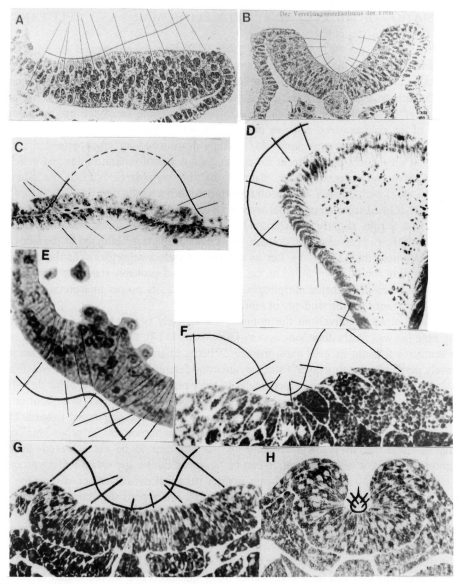

Figure 2.24. Some examples of the "prognostical" cell fans. A,B: neural rudiments of *Selachia* embryos, from Gurwitsch (1914). The author identified cell's axes with those of cell nuclei, the only ones visible by his staining techniques. C: initiation of a lateral bud in a hydroid polyp, *Obelia loveni*. A future contour of the bud is perpendicular to the diverged orientations of cell axes, most of all visible in the endodermal layer. D: an early tip of a hydroid polyp, *Dynamena pumila*. Its future configuration is perpendicular to the prolongations of the *distal* pieces of the bent cells axes. E,F: initiation of a blastoporal invagination in sea urchin's and in *Xenopus laevis* embryos, respectively. G,H: two successive stages of neurulation in the Urodelean (*Pleurodeles waltlii*) embryos. In all the cases the orientation of the cell axes 'predicts' further shape changes of a given rudiment.

Self-organization and tension dependence in the formation of cell domains.
The main problem associated with the formation of domains is whether their locations and dimensions are strictly predetermined (superimposed onto an epithelial layer in a one to one manner) by some previously formed structures (usually defined as inductors) or, even if the inductors are really required, they are no more than a kind of triggers, while the domains' localizing and domains' measuring capacities belong to as yet undifferentiated epithelial layer itself, as indicating its self-organizing properties.

In this respect, of a peculiar interest are the experiments made on hydroid polypes by Plickert (1980), who discovered that the domain's formation in these species stolons can be triggered by a mere pricking. So far as each domain gave rise to a new polype's bud, those were just buds rather than domains themselves to be traced. In the experiments described two successive prickings in the mutually different locations have been performed. It turned out that when the pricking sites have been located quite close to each other, they induced a common bud somewhere in between (Fig. 2.25 A). While being removed to a slightly greater distance, only the first pricking could induce a bud, while the second one did not (Fig. 2.25 B). Finally, while being removed even further, each one of the prickings induced its own bud (Fig. 2.25 C); the dimensions of the buds and, hence, of the initial domains' dimensions were roughly the same in all cases. Whatever may be the

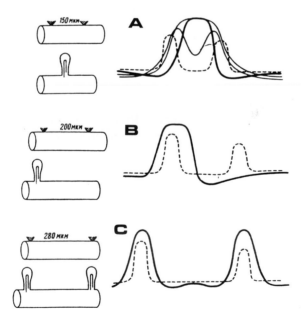

Figure 2.25. Induction of stolonial buds in hydroid polypes by two successive prickings performed in the different mutual locations. Left column: sketches of experimental results, right column: model constructions. A: the prickings removed from each other on a relatively short distance induce a single bud in between; B: if two prickings are removed to a greater distance only the first of them induces a bud; C: with a further distance increase, each one of the prickings induce its own bud. (From Belintzev, 1991, with the author's permission).

mechanisms of the buds induction, the experiments clearly indicate a lack of one to one relations between the location of the triggering agents and that of the created domains (buds), and hence the self-organizing capacities of a responding tissue. Later on we shall interpret these results in a more precise manner.

And now we shall address ourselves to another extensively studied rudiment, namely to the neural plate of amphibian embryos, which has also passed the domain stage in its development. It is well known that the specific differentiation of this rudiment into the neural tissue is associated with the induction process, the role of inductor being played by the underlining chordomesoderm. Does it mean, however, that the dimensions of an induced rudiment (including its transversal diameter) are determined by the inductor in a one to one manner, so that no signs of a self-organization within the neural plate itself can be traced? It was shown already long ago (Waddington and Yao, 1950) that an overall shape of a neural plate remained fairly normal even if the inductor material have been reversed to 180^0 around the vertical axis or dissociated into single cells. Such a result could not be obtained if a still non-differentiated neural plate did not possess considerable self-organizing capacities. The following recent data are also of interest in this respect (Zaraisky, 1991).

If removing from an amphibian embryo a part of its ventro-lateral ectoderm and mesoderm, which definitely do not contribute either to a neural plate or to a chordomesoderm, the dimensions of a neural plate will be nevertheless decreased proportionally to the diminishment of the total surface area of an embryo. At the same time, the amount of notochordal tissue have been decreased (and also only slightly) only if the operation was made no later than at early gastrula stage. This is in contrast with the behaviour of the neural tissue, its dimensions being proportionally decreased even if the removal of a non-neural area have been performed at the late neurula stage. These experiments make it obvious that the neural plate can effectively, and to a large extent independently from the underlaid chordomesodermal tissue adjust its dimensions to those of an entire embryo. In the other words, it reveals self-organizing capacities.

It was possible meanwhile to extend substantially (towards ventral) the neuroectodermal domain of *Xenopus laevis* embryos by a mere relaxation of the circumferential tensile stresses at the early gastrula stage (Beloussov *et al.*, 1990) - Fig. 2.26. Under the diminished tensions, the outlines of the domain appeared to be quite irregular (Fig. 2.26 B). Domains of columnarized cells could be also produced by a similar procedure in quite different and abnormal locations within embryos (Beloussov and Luchinskaia, 1995a). Combining all of the above data, we may conclude that the self-organizing capacities of the columnarization process are to a great extent depend of, and mediated by the tensions. We shall approach this point again in Chapter 3.

Contact cell polarization and the bistability of L/W ratio.
There are several evidences that a cell's transition from an initial, more or less isodiametric shape towards a columnarized one goes on as a kind of a contact-mediated relay phase transition, which we have already defined as a contact cell polarization (CCP). Such an idea is supported not only by the very continuity of the columnarized cell domains (Fig. 2.27 A), but also by the possibility of 'infecting' by a CCP relay the cells normally non-destined to become columnarized (Fig. 2.27 C,D). The average CCP rate in amphibian

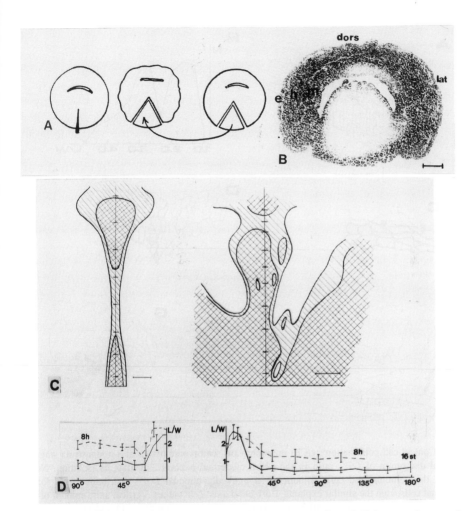

Figure 2.26. Experiments by relaxing tangential tensions on the surface of *X. laevis* embryos at the early gastrula stage. A: scheme of the operation. The tensions were relaxed by making a radial cut of a vegetal hemisphere and inserting into the wound gap a piece of a homologous endodermal tissue from another same stage embryo. B: a relaxed embryo in an hour after the operation, transversal view. Note an extensive ruffling of the relaxed embryonic surface and a thickening of the hypoectodermal cell layer. e: epiectoderm, h: hypoectoderm, m: mesoderm. C: two-dimensional maps of cell columnarization in the dorsal ectoderm of an intact embryo (at left) and of the same stage (8 h after the operation) relaxed embryo (at right). Vertical axis coincides with the dorsal embryo midline and horizontal axis corresponds to the blastopore level. Cross-hatched area corresponds to $L/W > 2$. Hatched area: $1.5 < L/W < 2$. Empty area: $L/W < 1.5$. D: diagramm of L/W ratios (vertical axis) for the epiectodermal cells (to the left) and hypoectodermal cells (to the right) of the same stage relaxed embryos (dotted lines) and the intact embryos (solid lines). Horizontal axes: angular distances of given cells from a dorsomedial line. As shown by C and D, in relaxed embryos, as compared with the normal ones, the extensively columnarized cell domains are much more extended towards ventral and have much more irregular shapes than in the intact embryos (From Beloussov et al., 1990).

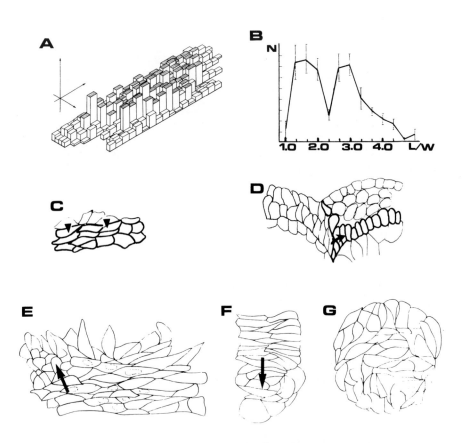

Figure 2.27. Contact cell polarization (CCP) in amphibians' embryonic tissues at a midgastrula stage. A: 2-dimensional diagramm of *L/W* ratio in the neuroectodermal explant 3 h after its isolation. Note a continuity of the columnarized cells domain. B: a markedly bimodal *L/W* distribution plot for the hypoectodermal cells from the similar explants. Horizontal axis: *L/W* values. Vertical axis: amount of cells (as a total, 467 cells belonging to 9 different explants have been measured). C, D: examples of a CCP transmission from the dorsal ectoderm of *X. laevis* embryos (light contours) to the cells of the transplanted ventral ectodermal pieces (dense contours). C is a tangential section and D is a transversal section. E: longitudinal CCP, F: transverse CCP, G: 'mixed' CCP in the tissues of intact *Rana temporaria* embryos. Arrows or pointers in C-F indicate the direction of CCP spreading. A, B from Petrov and Beloussov, 1984. C,D from Beloussov, 1980. E-G from Beloussov, 1978.

embryonic tissues has been estimated as 3 – 8 mkm/min (Beloussov and Petrov, 1984). In other words, about 5 min is required for involving each next cell into the CCP relay. Similar values can be borrowed from Keller's (1978) cinematographic data, related to bottle-shaped cells of a dorsal blastoporal lip.

CCP belongs to a large family of interrelated relay processes, to which also a 'contact induced cell spreading' in the cultures of pigment retina cells (Middleton, 1977), 'contact polarization of cell surfaces' in mammalian blastomeres (Johnson, 1981) as well as a relay of lobopodia formation in the chains of *Fundulus* blastomeres (Trinkaus, 1978) may be attributed. All of them imply obviously a coordinated set of the cytoskeletal and membranes transformations (Beloussov and Luchinskaia, 1983). Noteworthy, their rates fit those ascribed by Jaffe (1993) to slow Ca^{2+}–waves propagated by mechanical tension. A possible involvement of Ca^{2+} –signals in CCP spreading still deserves to be studied.

In embryonic tissues different geometrical forms of CCP can be distinguished. That one producing cell domains within epithelial tissues can be defined as a transverse CCP, since a relay is passing via transversal cell walls (Fig. 2.27 D,F). Besides, one can also observe a longitudinal CCP, which is going across vast cell masses, creating files of elongated cells (Fig. 2.27 C,E). The already described delamination process well may be considered as a kind of a rudimentary longitudinal CCP. Sometimes, a mixed version of CCP, intermediate between both the abovementioned can be traced (Fig. 2.27 G).

L/W distribution plots for the explants of amphibian embryonic tissues measured just during CCP process appeared to be markedly bimodal, showing the peaks under *L/W* = 1 and *L/W* ≥ 2.5, while the intermediate *L/W* values have been much less represented (Fig. 2.27 B). That permits us to suggest that a cell columnarization can be regarded, more or less tentatively, as a kind of a phase transition. Let us also recall in this respect, that the very transition from a non-columnarized to a fully columnarized state takes no more than few minutes, while the non-columnarized cell state may last for many hours and the fully columnarized state for no less than an hour.

A model of cell domains formation (Belintzev's model).
Now we would like to describe, in a very much simplified manner, a mathematical model (Belintzev *et al.*, 1985, 1987) that considers the formation of the domains as a kind of a self-organizing process. Paying a tribute to its first author, a theoretician Boris Belintzev (1953-1988) who died prematurely we will call it the Belintzev model (for its full Russian-language account see Belintzev, 1991).

The model is based upon the following empirical data:

(a) as has been just mentioned, there are good evidences for suggesting that a cell's shape is bistable. In the model's framework this is expressed by introducing within a model's equation a third order's term describing a non-liner 'point dynamics';

(b) the 'columnarization tendency' is spread from one adjacent cell to another owing to a CCP, which is considered, in the model's terms, as a kind of a diffusion-like process;

(c) cell columnarization is promoted by relaxation of the tangential tensions in a cell layer (Beloussov *et al.*, 1990; see 3.3.1 for more details).

All of the abovesaid can be formalized in the simplest and abstract, 1-dimensional case by the following equation:

$$\frac{\partial p_i}{\partial t} = f(p_i) + D_x \frac{\partial^2 p_i}{\partial x^2} - kT \qquad (19)$$

where
  p_i is a columnarization index of *i*-cell of a 'unidimensional' layer; this index is equal to the difference *L-W*, and not to *L/W* ratio, as previously;
  $f(p_i)$ is a non-linear (third order) term, describing the 'points dynamics';
  D_x is the diffusion coefficient, giving a measure of the CCP rate;

$\partial^2 p_i / \partial x^2$ is the second derivative of a cell columnarization index to the space coordinate, x; as we know already from Chapter 1 (p. 56), this is an indispensable component of the diffusion term;

k is a phenomenological coefficient;

T is an elastic force, up to now of an indefinite origin ($T>0$ corresponds to stretching, while $T<0$ to compression).

In its present form, meanwhile, the equation *(19)* does not describe a *self*-organization, since it implies an external mechanical force of an unknown origin, expressed by the term kT. Only as soon as we shall formulate a way, by which this force might be generated within the layer itself, owing to the columnarization of its cells, the equation could be defined as a 'closed' one, that is, capable of describing the real self-organizing properties of a layer. To make this possible, it is enough to introduce very simple boundary conditions. Namely, one should suggest that the layer's edges are fixed, that is, the layer is incapable to reduce its length as a result to the cells columnarization. On of the most natural ways to make this is to suggest that the layer is encircling a cavity with a fixed volume. If this takes place, a progressive cell columnarization will inevitably increase the tangential stretching (tensile) force along the entire layer. If also taking now into consideration a premise (c) of the model (inhibition of further cell columnarization by the increased tension), we shall get a long-range negative feedback between the tangential tension and cell columnarization, acting throughout an entire layer. Together with a short-range positive feedback, exemplified by CCP (stimulation of a columnarization in the adjacent cells), that gives a closed construction which is endowed by the self-organizing properties an is formally similar to the above discussed OMH model (p.113-115). Meanwhile, further elaboration of Belintzev's model will bring us towards some new important conclusions.

For describing, in the model terms, the tension generated by a columnarization of cells within a layer with fixed edges, it is reasonable to introduce the notion of an *average* cell columnarization $\langle p \rangle$ throughout an entire layer, which consists of n cells:

$$\langle p \rangle = \frac{\Sigma^n p_i}{n}$$

The tensile force generated by the average cell columnarization is

$$T = k \langle p \rangle$$

The next question will be, how do this force affect further columnarization of the individual layer's cells. It is reasonable to assume, that the effect will be negative for the cells having a columnarization index less than $\langle p \rangle$ and positive for the cells with a columnarization index greater than $\langle p \rangle$. This idea is expressed by introducing within a model a negative member

$$-k\varepsilon (\langle p \rangle - p_i) .$$

The biological meaning of the coefficient ε is worth discussing. Formally, it describes an active cell's response (its ability for further columnarzation) as a function of the difference between the average cell columnarization throughout a whole layer and the momentous columnarization of a given cell. In more general terms that means that it relates the active cell's response to an overall (averaged throughout a whole layer) passive tensile force, and, also, to a momentous cell state. This is, to my knowledge, a first and probably a sole attempt to formalize the idea of the active-passive stress feedback between the events belonging to different levels.

After these additions the resulting equation will look like

$$\frac{\partial p_i}{\partial t} = f(p_i) + D_x \frac{\partial^2 p_i}{\partial x^2} - k\varepsilon\left(\langle p \rangle - p_i\right) - kT, \qquad (19a)$$

its last term expressing an external elastic stretching force of an arbitrary origin. In the absence of this term, (19a) describes now an epithelial layer as a mechanically closed system, all of the acting forces being generated within.

As shown by a corresponding analysis of the model equation (for more details see Belintzev *et al.*, 1987) after $|k\varepsilon|$ passing a certain threshold (this threshold value depending upon the stability of the cell's isotropic state and on the value of the diffusion coefficient D_x) an initial homogeneous (domainless) state of a cell sheet becomes firstly metastable and then, with a further rise in $|k\varepsilon|$, completely unstable. Correspondingly, the domain(s) will appear at first as a result of a finite, and later on of an infinitesimal, perturbation. The following properties of the formation of domains are important most of all:

(1) If a cell layer does not show any elastic reaction at all (its edges are not fixed) the length of a columnarized cell domain will be indefinite (Fig. 2.28 A), being changed after any next polarizing perturbation. In the other words, the layer's segregation will in this case be non-robust. Meanwhile, the robustness will immediately appear as soon as the layer exerts an elastic reaction to a deforming force. Under the latter conditions the following situations deserve to be distinguished.

(2) There is initially a monotonous gradient of a cell columnarization along a layer. Now, if the model is at work, the gradient will become more abrupt, so that a layer will become stably segregated into two domains with fixed proportions (Fig. 2.28 B).

(3) Even in the absence of such a gradient (cell columnarization index is initially the same throughout the entire layer) any local "polarizing perturbation" (like pricking in the above described Plickert's experiments) will be enough for initiating a creation of a columnarized domain with its final length being strictly determined by the model's parameters (Fig. 2.28 C).

(4) Just the same will take place if two such perturbations will be situated closely enough to each other (Fig. 2.28 D).

(5) However, if both perturbations are localized at the opposite ends of a layer, each one of them will give rise to its own columnarized cell domain (Fig. 2.28 E).

It is easy to see that the results illustrated by Fig. 2.28 C-E correspond to those of the abovementioned experiments in hydroid polyps (Fig. 2.25 A-C).

A remarkable property of the Belintzev's model is that the ratios of the lengths of the

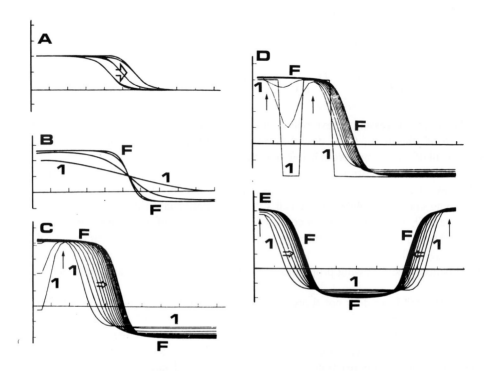

Figure 2.28. Formation of columnarized cells domains according to the Belintzev's model. A: in the absence of an elastic reaction of a cell layer the domain's length is indefinite. All the other cases imply the elastic reaction. B: formation of the domain while starting from a smooth gradient of a cell columnarization. C: same under the influence of a single initial polarizing perturbation. D: same, two closely situated perturbations. E: same, perturbations situated at the opposite ends of a layer. In B-E 1 is the initial state and F is the final state. Vertical arrows indicate initial perturbations while horizontal arrows point to the direction of the columnarized domain's extension. Horizontal axis: length of a layer, vertical axis: index of a cell columnarization (L-W). From Belintzev, 1991.

columnarized   cell domains   to   the flattened parts of   a sheet are, for the fixed parameters'values, always the same, irrespectively of the absolute total length of a layer. In the other words (quite unexpectedly, and without any additional assumptions!) this theoretical construction reproduces Driesch's embryonic regulations, which is not an easy task for any model. Its correlation with the Driesch's conclusions, as we exposed them in

Chapter 1 (pp. 35–37) goes even further: the model shows a way for a really *delocalized* regulation of morphogenetic processes, without each cell 'knowing' its exact position in relation, say, to the edges of a layer. Actually, for participating or not participating in the formation of domains, a cell should 'know' nothing except two things: its own columnarization index and the value of the tangential tensile stress, averaged throughout the whole layer and determined, for the case of its mechanical closeness, by $\langle p \rangle$. The first 'knowledge' is local, while the second one is global, but none of those contain any kind of a 'positional information' in its strict sense. The local perturbations also cannot be considered as the sources of such an information, because, as we can see, the relations between the locations of the perturbations and of the domains produced are quite far from being unambiguous. In the model's terms *it is a discrepancy between the local shape of a given cell and the global mechanical properties of an entire layer which moves its morphogenesis forward*: the global shifts initiate the local responses, the latter changing in their turn the global properties, and so on. Feedbacks between the local and the global (that is, between the different levels' events) are here quite obvious. This is, by my view, the most attractive property of the Belintzev' model, deserving further elaboration.

2.3.6. FROM FANS TO FOLDS: A 'QUASIRELAXATION' STRATEGY

As we have already mentioned, those columnarized cell domains which are destined to create a fold sooner or later take a fan-like configuration. Any fan is formed as a result of a CCP relay, implying quite an active behaviour of the recruited cells. Such an activity follows directly from the fact that each next cell to the side of a fan is inclined towards a fan's center to a greater extent than its medial neighbour (Fig. 2.24 C, E-G). That cannot be achieved without a substantial resorbtion of the apical cell surfaces and an unequal stretching of their transverse surfaces. As a result, the fan's cells become highly asymmetrical, their asymmetry and, hence, S/V ratio considerably increasing towards a fan's periphery. It is already this asymmetry which points to a highly tensed state of the cells composing a fan. Same conclusion may be drawn from the abovementioned (p. 78–79, Fig. 2.1.) dissection experiments.

The matter is, that most of the dissected tissue pieces consisted of cell fans, or were the parts of such fans. This is the case, among others, for the blastoporal area of an early gastrula embryo and for the neural plate of a neurula stage embryo. If, now, detaching a cell fan from an underlaid tissue in a latero–medial direction (towards a fan's center), it curls immediately (so to say, 'under the knife') in the same direction as a similar embryo region does so, although in a very much retarded rate, during its normal development. For example, a blastoporal area imitates a closure of the blastopore, the neural plate imitates the neurulation folding, and so on (see Beloussov *et al.*, 1975, for more details). We have already presented the arguments which support the idea that these fast deformations can be regarded as the passive relaxations of the pre-existing stresses (those stored within the highly oblique fan's cells). Such a conclusion is supported by the fact that during these fast bendings a cell shape is changed from a highly oblique towards a more symmetrical, and even isodiametric shape. That indicates the overall contraction of the cell surface, that is a minimization of the previously (during cell's obliquening) increased S/V ratio. Remarkably, the cell shape changes during normal morphogenesis of the fan-forming epithelia show just the same tendency: the percent of the oblique cells is diminished while that of the more symmetrical (quasi-rectangular) cells increases (Fig. 2.29). Obviously, it is this tendency of the oblique fan's cells to increase again their symmetry order, which is the basis of a fan's 'prognostication', firstly noticed by Gurwitsch.

In such a way we come to a conclusion which looks at first glance paradoxical: both the fast relaxations and the normal 'slow' morphogenetic processes are associated with the same kind of cell shape changes, and are going, in this respect, along the same pathway. Certainly, that in no way means that under normal conditions the folding epithelia do not overcome the mechanical resistance of the surrounding tissues and do not consume energy: they definitely do this, and in some cases this energy can be roughly measured (see Bereiter-Hahn, 1987). However, the deformations of a layer's cells (determining that of an entire layer) are such as if they move towards a mechanical equilibrium; we may define this tendency as a *quasirelaxation*. It may be compared with the movement of a body down a slope with a very high friction (say, along wet snow); the energy should be spent indeed, but it has no need to be vectorized: the direction of the movement is already determined

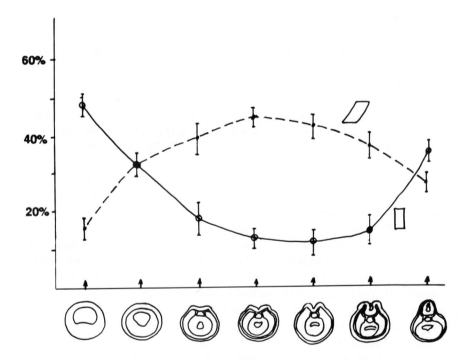

Figure 2.29. A percent of the obliqued (dotted line) and the rectangular (solid line) cells in the neural plate of *Rana temporaria* embryos at the developmental stages diagrammed to below. Courtesy of V.G.Cherdantzev.

by the slope's inclination. An advantage of such a mode of behaviour is obvious – you should not be anxious about precisely vectoring and/or quantifying a motive force (anyone who ever hammered a nail into a desk knows how difficult it is sometimes). Instead of doing this, for moving a sledge down a viscous slope it is quite enough to lift it a bit or simply shake it. With such a strategy in hands, an epithelial folding acquires a desirable robustness.

Another advantage of a 'quasirelaxation' strategy is in providing additional possibilities for determining the dimensions (transversal diameters) of the folds. This can be done by modifying the time duration between the start of a fan's formation and the beginning of its quasirelaxation, this latter directly moulding a fold: obviously, the less is this time interval, the narrower the fan and, hence, the fold would be. On the other hand, a substantial time delay between both processes will give rise to very extensive folds. Such transformations of the temporal shifts into spatial patterns seem to be widespread in development, not in the last turn because of their accessibility to the relatively simple regulatorial factors, which do not need to be precisely arranged in space.

It is worth mentioning, that such a biologically plausible strategy is practically un-achievable, if treating a folding process as a purely passive one, as it would appear in non-living bodies. If considering, for example, an epithelial layer simply as an incompressible solid body with elastic properties (as it actually is from the point of view of a 'passive' mechanics) one would not be able to explain why a contraction of one of its surfaces would not lead immediately (with an elastic wave rate) to the corresponding bending of an entire layer. And if (as is actually the case) such an immediate bending does not take place, it becomes even less clear why the same bending, following a definitely predicted pattern, will regularly take place some time later.

The revealing, by an embryonic epithelial layer, of just this mode of behaviour means that the layer is really a living being able, firstly, to perceive an imposed mechanical situation and, secondly, after some time delay, to elaborate a kind of a 'reasonable' active response to it. What is actually perceived is, in our case, the mechanical stress pattern embedded within a fan, while a somehow delayed reaction to it is associated with the quasirelaxation of this stress.

Thus, a fan's strategy is markedly biphasic: firstly a stressed configuration, storing in it all of the morphological characters of a future fold (location, dimensions, and shape) is established and only later on, after a definite delay, these previously hidden properties became visualized. The very essense of such a strategy, including its specifically 'vital' character was clear already for Gurwitsch, who defined the second phase of the process as being 'dynamically preformed' in the first. Translating this statement into more ubiquitous (but, probably, in this case less precise) modern terms we may claim, that all the 'information' about a fold's properties is already contained within a fan. In any case, what is most important of all, such a preformation (or the information storage) is of a markedly holistic character: it is a fan as a whole, rather than its individual cells, which contains it. Recall, that according to the Belintzev's model the cell domains (including fans) are 'dynamically preformed' within an entire cell sheet in the same holistic and largely delocalized way. As a result we obtain, although if up to now in quite general outlines, a system for holistic determination of epithelial morphogenesis. Note that such a system includes in it an entire hierarchy which consists of few discrete steps, each one of them having its own L_{ch} and T_{ch}. The hierarchically coupled L_{ch}-s are, in descending order: a maximal interaction distance between two just formed neighbouring domains (for example, that between the buds shown in Fig. 2.25 B, C) and a length of each single domain, taken separately. The associated T_{ch}-s are: the duration of a domain's formation and the duration of its 'quasirelaxation' into a fold. Both latter ones are, as a rule, of the range of about several dozens minutes.

The epithelial folding is not the only possible result of a quasirelaxation of a fan. Obviously, the folding would take place if and only if the resistance of the surrounding structures to the deformation of a layer is not too great (Fig. 2.30 A). Otherwise one might imagine that the symmetrization of the obliqued fan's cells would better lead to the antiparallel mutual shifts its surfaces (Fig. 2.30 B). Just this situation will be discussed in the next section. A quasirelaxational dynamics can also play an important role in the previously described delamination and cavitation events. Firstly, a symmetrization of the oblique cells which are concentrated, as mentioned above, at the ends of an already delaminated area (see Fig. 2.20 C,D), would promote further straightening of the delamination furrow at both sides of the oblique cells (Fig. 2.30 C). And secondly, a symmetrization of the cells of two adjacent mirror–symmetrical fans will form a cavity in between (Fig. 2.30 D). Note that both the location and the dimensions of the structures to be arised should be determined by tensile patterns established in their neighbourhood.

In concluding this section let us trace the changes in symmetry orders at the different structural levels. As mentioned above, at the cell fan stage there are individual cells which are highly asymmetrical while the entire fan is, as a rule, symmetrical. After folding the

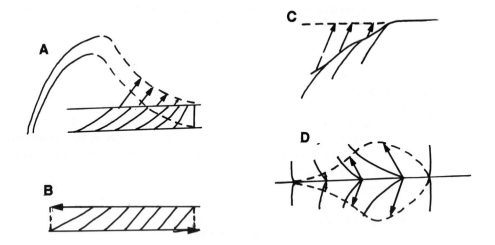

Figure 2.30. Different morphological consequences of a "rectangularization' of the initially obliqued cells. A: bending of a cell layer. B: antiparallel shifts of the opposite surfaces of a layer. C: straightening of a layer's border or of a delamination line. D: formation of a cavity. In all the cases the final configurations are shown by dotted lines.

situation becomes reversed: cells approach a more symmetrical shape, while a layer as a whole reduces its symmetry order, transforming into a fold quite often having a complicated geometry. This is one of the many examples of a 'symmetry exchange' between the different structural levels.

2.3.7. LINEAR EXTENSION AND SHAPING BY MEANS OF GROWTH PULSATIONS: A HYDRAULIC MORPHOGENETIC MACHINE, EMPLOYING A FAMILIAR STRATEGY

Now we come to a rather peculiar mode of epithelial morphogenesis which is expressed most clearly in the vegetal buds of hydroid polyps (although similar processes take place also within sponges, blastoderms of bone fish' embryos, chicken embryos and even in growing algae – Beloussov et al., 1993). All of these processes are characterized by a very regular periodicity of several minutes. Thus, in hydroid polyps the growing rudiments are regularly extended and retracted within species-specific periods, ranging from 1 min (a fresh water nerve-free *Hydra attenuata*) up to about 14 min (a marine hydroid polyp *Dynamena pumila*). Actually, a space–temporal pattern of these periodical events can be rather complicated (Figs. 2.31, 3.18, 3.19).

These events, which we call growth pulsations (GP), are based upon quite remarkable cell activities, combining vacuolar dynamics similar to that observed in Protista with definitely cooperative cell reactions typical for Metazoans. The resulting type of morphogenesis is, probably, among the evolutionarily most ancient ones, and looks, so to say, straightforwardly mechanistic. However, it is quite precise and perfectly regulated at the different levels. Let us trace it in more details, taking as the main examples the marine colonial hydroids of the *Obelia* genus. We shall start from describing their histological structure.

The so called coenosarc, that is a living tissue of hydroid polypes is known to consist of two adjacent sheets of cells, the ectoderm and endoderm. Both layers are formed mainly by the epithelio-muscular cells, which are at the same time the only active actors in a GP scenario. In the first approximation both layers participate in GP in a similar manner (actually it is not so, but for our immediate purposes such a rough picture is appropriate; we shall postpone a more detailed analysis up to the Chapter 3). That permits us to restrict our description up to now to the ectodermal layer only. It is covered by a chitinous envelope, the perisarc, which is secreted at the very tip of a growing stem as a thin viscous layer, to be later on rapidly (within few minutes) hardened. Thus, in a growing stem a disto-proximal gradient of a perisarc hardening is established, the more to the proximal the harder is the perisarc. An entire bud morphogenesis may be regarded as a moulding of a still soft distalmost perisarc by the mechanical activity of the underlaid cells. A hardened perisarc perfectly fixes ('memorizes') the achieved shapes. Each single growing stem has a distal zone of several dozens cells in height (these cells being firmly adhered to the inner perisarc surface by their apical surfaces) and an indefinitely long (extended up to a colony base) proximal zone, its ectodermal cells being detached from the perisarc and spread over the mesoglea. Only the distal zone cells are involved in GP. The morphogenetic activities associated with the more proximal zone will be discussed later on (see Chapter 3).

As it can be derived from the time lapse pictures, complemented by histological data, during each next extension GP phase the cells of the distal zone change their orientations from more oblique towards more transverse ones, these reorientations going on in a proximo-distal succession (Fig. 2.31 A–C). The rate of such a proximo-distal wave of the cells' involvement into transverse rotations is about 40 – 90 mkm/min. In *Obelia loveni* stems the transverse rotations take about 3 min out of a total 5 min GP period. During rotations the apical cell borders remain firmly adhered to the inner perisarc surface, while their basal ends are shifted distally up to 25 – 35 mkm, pulling the mesoglea from below together with a prolonged column of the attached cells (Fig. 2.31 A, B, upwards pointing vertical arrows). At the same time the rotating cells push the still soft distalmost perisarc, lifting it up to 5 – 10 mkm in each GP extension period; this amount is, as we can see, considerably smaller than the abovementioned shifts of the basal cell ends. The existence of such a difference means that during their transverse rotations the more distal cells are compressed from below by their proximal neighbours. Such a cell–cell compression seems to be the main physical factor moulding the shapes of the hydroid buds.

At the extension GP phase a rudiment not only elongates itself, but also increases its own relief: any curvature differences on its surface, even if beforehand quite slightly outlined, are now increased while being smoothed again (although if not exactly towards the starting point) at the next retraction phase (Fig. 2.31 E, F).

In about 1 min after the cessation of transverse rotations wave the distal zone cells return, almost simultaneously, to their initial oblique positions and cease to stretch the uppermost perisarc (Fig. 2.31 D). A stem is shortened, obviously owing to the elasticity of both the mesoglea and still non-hardened perisarc: this is the retraction GP phase. However, the amount of the retraction is, as a rule, less than that of the previous extension, since each successive growth advance is partially fixed by the very geometry of the distal perisarc: it is much easier to push a cusp out, than to pull it back. In such a way a net progressive elongation of a rudiment is going on.

The most remarkable property of the retraction phase is a rapid (about 1 mkm/s) and almost synchronous distalwards sliding of the disto-apical cell poles along the inner surface of the perisarc; this is associated with the formation of the prolonged distally-oriented 'necks' (Fig. 2.31D, 2.32A, 2.33 B). Because of the sliding the cells shift their fixation points distalwards along the inner perisarc surface, for entering later on a new round of transverse rotations.

What are the forces responsible for the regular transverse cell rotations? As shown by electron microscopy, at the height of the extension phase the cells become full of isolated spherical vacuoles, which are obviously under a considerable turgor pressure (Fig. 2.32 B, 2.33 A). On the other hand, during the retraction phase the vacuoles become fused into prolonged channels, opened in the extracellular space (Fig. 2.32A, 2.33B). As demonstrated by X-ray elemental analysis (Kazakova *et al.*, 1994), ionic content of the extension phase vacuoles considerably differs from that of the external environment, while at the retraction phase the differences are to a large extent smoothed out. All of these observations are consistent with the idea that at the extension phase the vacuoles are indeed the isolated compartments which exert a substantial osmotically-driven turgor pressure. It is this pressure to be transformed, within a cell layer, into the above-

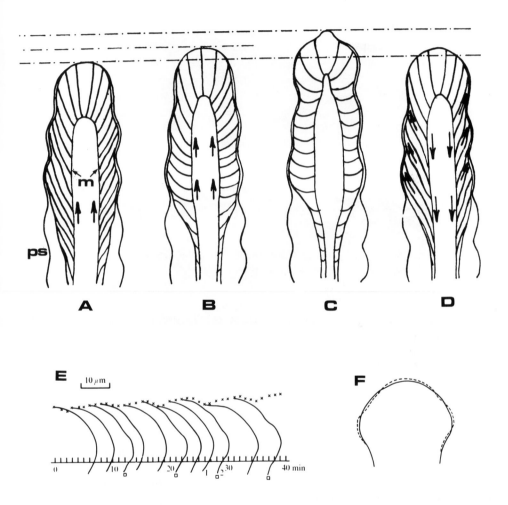

Figure 2.31. Cell activities and morphological events associated with growth pulsations (GP) in hydroid polypes: a summary of histological and time-lapse studies. A-D: schematic representation of cell shifts and reorientations during single growth pulsation within a growing tip Horizontal lines at the top indicate distoproximal levels reached by the tip at the different phases of GP. Only ectodermal layer, perisarc (*ps*) and mesoglea (*m*) are shown. A-C frames corresponds to the extension phase while frame D to the retraction phase. E: changes in the outlines of an *Obelia* hydranth bud during several successive GPs. At the height of extension phases the curvature differences (the reliefness) of a hydranth bud's surface are maximal. F is a superposition of the bud contours at the retraction (solid line) and the extension (dotted line) phase of the same GP.

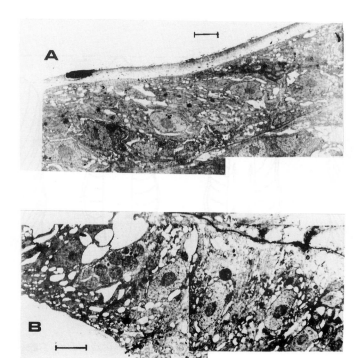

Figure 2.32. Electron microscopy view of *Obelia* ectodermal cells at the retraction (A) and the extension (B) phase of a growth pulsation. In the both cases perisarc is to the above. Note a series of the distally oriented narrow cell "necks" and the prolonged vacuolar channels in A, as opposed to the numerous rounded vacuoles shown in B. From Beloussov et al., 1989.

mentioned cell–cell compression force (Fig. 2.33 A, short vertical arrows). On the other hand, at the retraction phase the pressure is released and cells are firstly relaxed and then actively stretched along their longitudinal axes (Fig. 2.33 B).

How are the osmotically-generated pressure stresses transformed into cell rotations, extension of the rudiments, and their shape changes? The first step on this way is the trend of a body, covered by an elastic shell, to approach, as the internal pressure increases, towards a more symmetrical shape. For tightly packed epithelial cells, firmly connected with their neighbours, the most symmetrical shape among those possible is that of a rectangle. The hydroids' cells rapidly come indeed to this very shape after a relaxational dissection of a rudiment. In such a way we come to the familiar situation of a quasirelaxational shape change, the additional (non-vectored) force being of osmotic origin. At the level of the entire layer these cell deformations will produce the following morphogenetic results.

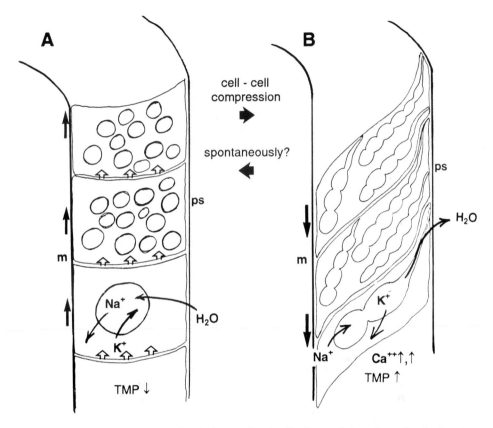

Figure 2.33. A sketch of mechanochemical events involved in the regulation of growth pulsations in hydroid polypes. *m*- mesoglea, *ps* - perisarc. For other designations and comments see text.

Firstly, in the regions situated proximally to the growing tip, and thus covered by a hard non-deformable perisarc, the only possible result will be a linear extension of the rudiment as a result of the distalwards traction of the mesoglea–attached cells (Fig. 2.31 A,B, 2.33 A, solid arrows). Meanwhile, the distalmost regions covered by a thin just deposited perisarc can be really deformed by a cell–cell pressure. This is evidenced by the increase of the local curvatures differences just at the extension GP phase (Fig. 2.31 E, F), that is, at the moments of the highest cell–cell pressure. To the first approximation, this may be regarded as a passive deformation of a pressurized shell with its edges fixed. As a very mechanistic imitation of this process we can take a heated rail, relaxing the internal compression stresses by buckling further its areas that initially have the maximal curvature. The resulting displacements will be directed radially, outwards from the center of a curvature, and be roughly proportional to the local curvature, that is (in the unidimensional approximation) to $1/R$, where R is the radius of the local curvature, that is a curvature measured in a vicinity of a given layer's point (Fig. 2.34 A). However, in developing systems such a purely passive mechanical component will be almost certainly supplemented by an active one, going with the same direction. In other words, we should deal here again with a quasirelaxation, now of the compression stresses. In this particular case we may use a simple model of epithelial folding (Beloussov and Lakirev, 1991), which can be defined as a 'curvature increasing rule': any slight curvature difference tends to increase itself, automatically producing on its flanks, even if they were initially flat, new maxima of the local curvatures of an opposite sign. If also introducing into the model the different coefficients of a layer's elastic resistance to the curvature–dependent deformations (Fig. 2.34 B) and the different ratios of the durations of extension and retraction GP phases we obtain a set of shapes shown in Fig. 2.34 C. This set is, on the one hand, rich enough, and, on the other hand, concentrated around a single 'archetype'. In a planar projection the archetype has a shape of a trefoil. If introducing the third dimension, it may either remain to be a flattened trefoil, or become a rotational body. We will discuss this set of the generated morphologies again in the Chapter 3 (mostly in 3.4.4). What deserves to be emphasized here is that the resulting shapes can in no way be directly deduced from those monotonic swinging processes which are taking place within individual cells and the entire cell column. Instead, the shapes are generated entirely at the whole bud level (as depending upon its geometry), being in no way imprinted in the individual cells' activities *per se*. In parallel with what has been said about cell fans, we may consider an entire, still smooth (but definitely shaped) contour of an as yet non-differentiated rudiment as containing the 'morphogenetical information' about its further shape, or as 'dynamically preforming' it, by Gurwitsch's terminology. In any case, to make this possible the rudiment must be mechanically stressed.

Let us now pass, as far as the data available permits it, to the problem of a chemomechanical transduction during GP. This implies two interrelated questions: (1) how the energy of a cell's metabolism is transformed into the turgor pressure within cell vacuoles; (2) what are the molecular (or, more vaguely, cytophysiological) factors providing a regular switching from an extension towards a retraction phase and back again, generating thus a regular pulsation regime (a problem of a signalling). At the present time only quite provisional answers to the both questions may be given.

The first question can be elucidated to some extent by the already mentioned results of X-ray elemental analysis and the measurments of the transmembrane potentials (TMP) (Kazakova *et al.*, 1994 and in preparation). It turns out that during the extension phase the potassium/sodium ratio within vacuoles largely exceeds that of the surrounding sea water, while for the retraction phase the reverse is true. On the other hand the TMP is considerably decreased at the end of extension phase and restored during retraction phase (Fig. 2.33, cf. A and B). To this we may add that ouabain, the inhibitor of Na^+,K^+ ATPase, prolongs the extension phase, while tetraethylammonium, the blockator of potassium channels, induces the retraction phase (Beloussov *et al.*, 1989) As a result, we come to the following suggestion: (1) the extension GP phase is associated with the inhibition of the Na^+,K^+ ATPase in the vacuolar membranes; (2) as a result, potassium is leaked out of the cytoplasm into the isolated vacuolar space and becomes (together with some as yet undetected anion) the main osmogenic ion, providing water inflow and

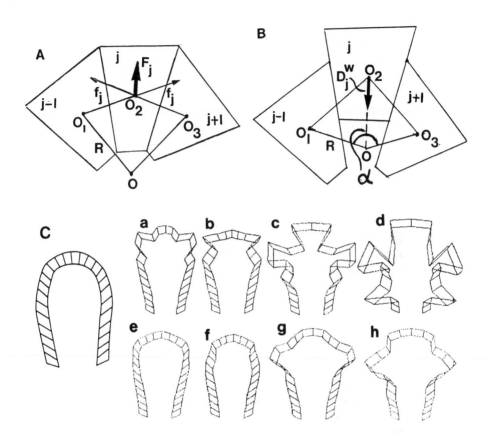

Figure 2.34. Finite elements modelling of the morphogenesis of tubular rudiments exerting an internal pressure. A, B: models algorithms. In A the main algorithm of the 'curvature increasing rule' is imaged, the resulted centrifugal force F_j arising from the vector addition of the lateral pressure forces exerted by j+1 and j-1 cells. B illustrates the elastoviscosity force D_j^w, as opposing to F_j and tending to smooth out any curvature increase. Panel C gives the results of modelling each time starting from the shape shown to the left. Among the resulted shapes those comprising the upper row correspond to the smaller elastoviscosity coefficient rather than those comprising the lower row. Each row members (a-d and e-f) are arranged from left to right accordingly to the increase of their extension/retraction durations ratios per each modelled GP. Thus, for a, e the duration of the extension phase is the smallest, while d, h shapes are modelled under the assumption of a permanent extension, that is the duration of the retraction phase equals to zero (From Beloussov and Lakirev, 1991).

hence increase in the turgor pressure. At the same time sodium enters the cytoplasm so that the sodium/potassium ratio within cytoplasm approaches that in the external environment, providing TMP decrease. (3) during the retraction phase the activity of the Na^+,K^+ pump is rapidly restored, causing the removal of the excessive sodium from the cytoplasm to the vacuolar compartment (at the expense of potassium). Within such a framework it well may be just the TMP energy that is transformed during the extension GP phase into the intravacuolar osmotic potential. We can see that the mechanical 'working stroke' of a cell is associated with the discharge of its chemical energy, as it should take place in any mechanochemical machine. Let us also note that this discharge is associated with an increase in the symmetry order of the individual cells (which are moved towards rectangular shapes) and, at the same time, with the decrease of a symmetry order of an overall rudiment (as indicated by its folding). On the other hand, the retraction GP phase (that of cell's dissymmetrization) is associated with the recovery of the electrochemical energy of the individual cells (this energy being stored in TMP), dissymmetrization at the single cell level and increase of a symmetry order at the entire rudiment's level (as shown by its contours smoothing). The 'exchange of the symmetry orders' between the adjacent levels is here quite obvious again.

Approaching now a signalling problem, we may first of all definitely conclude that there are no autonomous oscillatory clocks within single cells. The cells exert oscillations only while being the members of a densely packed (pressurized) sheet closely adhered to the perisarc by its apical surface, and immediately stop oscillating if only one of the above mentioned conditions is absent: for example, if they are deadhered from the perisarc or if the pressure within a sheet is relaxed. And even if all of the above conditions are provided, the cells can be easily blocked for a substantial time period (although no more than for several dozens of minutes) either at the extension or retraction phase. The first of these responses requires a hypotonical external medium (providing an increased turgor pressure), while the second a hypertonical medium (which decreases the turgor pressure). The possibility of both the extension and the retraction cell states being maintained within time periods, largely exceeding those of the corresponding pulsations phases, points to the existence within cells of some stabilizing, that is, positive turgor–dependent feedbacks. In mathematical terms (see Chapter 1, pp.51-53), one may tell that the single cells are working in the waiting, rather than the autooscilalting regimes. But where are, now, the negative feedbacks, providing the GPs going on with their specific rates, coming from?

As has already been mentioned, a normal passage towards the retraction GP phase is preceeded, for each cell of a layer, by a substantial pressure, exerted by its proximal transversely rotated neighbour. Such a pressure can be easily imitated by bending a stem artificially: under these circumstances it becomes applied to the concave cell layer of a bent stem. In 2–3 minutes after bending (which fits the characteristic GP time) this layer's cells take a structure typical for the retraction phase (Kazakova *et al.*, 1997). We suggest therefore, that a cell–cell pressure should be among the main signals triggering a passage towards a retraction phase. As to the possible molecular mediators of the signal(s), the $[Ca^{2+}]_{cyt}$ dynamics may be regarded at the moment as a plausible candidate. Two brief successive increases of $[Ca^{2+}]_{cyt}$ could be indicated indeed at the beginning of the retraction phase (Kazakova *et al.*, 1995). Probably, it is this $[Ca^{2+}]_{cyt}$ increase which induces the vacuoles' fusion into the opened channels and their subsequent contraction, typical of the retraction GP phase. We do not know up to now, meanwhile, whether the opposite transition from the retraction to the extension phase requires a special signal, or whether it is to a greater extent a spontaneous process (as being associated with an electrochemichal energy discharge).   A more or less coherent and biologically substantiated mathematical model of the growth pulsations is still lacking. It is highly probable, meanwhile, that: (1) the pressure stresses and the Ca^{2+}-dynamics should be its main components; and (2) that it should imply the parameters' drift with a rate comparable with that of the dynamical variables: this is indicated by the very asymmetry of the GP diagramms. Correspondingly, the models of GP cycles may be similar to those shown in Fig. 1.19 C, D. In particular, we will get something similar to Fig. 1.19C if plotting a cell–cell pressure along X axis and a reverse $[Ca^{2+}]_{cyt}$ along Y axis. On the other hand, Fig. 1.19D recall us a typical time record of a GP (see also Fig. 3.20)

This is, probably, a proper point to cancel (for the moment) the story about growth pulsations, actually a sad story about the monotonous life of several dozens of cells, enslaved by the 'whole' and even having within themselves no image of the perfect shapes which they collectively create.

2.4. Stress-mediated morphogenetical feedbacks: attempts of generalization.

2.4.1. FORMULATION OF A STRESS HYPERRESTORATION HYPOTHESIS

Two simple interconnected conclusions can be derived from what has been said in this chapter: (1) all of the morphogenetic processes, irrespectively of their scale, are associated with pronounced and regular space-temporal ups and downs in the mechanical stress values within the cells and tissues involved: for example, the periods of the tensions increase are followed by their relaxations, which as a rule are accompanied by a rise in tensions in the neighbouring parts, etc.; (2) all of the stress producing devices, be they belonging to a supramolecular or to a supracellular level, are mechanosensitive, that is, are able to perceive stresses and to react onto them. This is enough for suggesting the very simple idea that the feedbacks between the passive stresses and the active stress generating devices may be of a real morphogenetic importance and be even regarded as the 'driving forces' of the morphogenesis.

Such an approach is not completely new. The first paper applying this idea to the embryonic morphogenesis belonged, by our knowledge, to Odell *et al.* (1980). As we have briefly mentioned beforehand, the authors postulated the existence of '+, +' feedback between the active contraction and passive stretching within an epithelial layer. This idea have been criticized by Belintzev *et al* (1987) as being unable to reproduce a differentiation of an initially homogeneous cell layer. Instead, as we know, the latter authors ascribed a primary importance to the '+, –' feedback between the active cell's columnarization (=tangential contraction) and the passive tangential tension, as providing a regular segregation of a layer into the domains of columnarized and stretched cells.

This does not mean, however, that Odell *et al.* model do not reflect at all any real morphogenetic situations: these ones fitting the model's assumptions will be described below. On the other hand, Belintzev's model, inspite of interpreting nicely some important aspects of epithelial morphogenesis, also cannot be considered as universal. Firstly, it accepts some specific border and initial conditions as given. Secondly, it cannot be effectively applied to the morphogenetic processes associated with an extensive cell's repacking, such as those involving the immigration or intercalation of cells. We know, meanwhile, that the latter processes are morphogenetically very important and stress–dependent ones.

Would it be possible under these circumstances, to formulate some more general concepts of the feedback relations between active and passive stresses, including the situations modelled by both Odell *et al.* and Belintzev *et al.* as particular cases? In discussing these topics in 1990, my friend Jay Mittenthal of Illinois University, Urbana, joined me in suggesting that at the present time any such generalization can only be formulated as a kind of more or less vague qualitative statement, centered around the biological meaning of a given reaction. From such a point of view, the following is obvious: (1) the feedback relations should be evolutionarily ancient, since the primitive organisms have already been confronted with the mechanical forces and the consequences of their action; (2) they should generally be homeostatic, that is directed towards the restoration of the initial, 'habitual' stress values and against any dangerous deviations from these values; (3) they should be robust, since it would be 'cheaper' and more reliable to produce some overshoots in stress restoration, rather than to develop any highly precise compensatory responses.

Adding practically nothing to these almost trivial ideas, we suggested the following generalization, defined as a 'hyperrestoration rule' (HRR):

A cell or a piece of tissue, after being shifted by an external force (either introduced artificially, or exerted by another part of the same normally developing embryo) from its initial stress value, develops an active mechanical response, which is directed towards restoring this stress value, but as a rule overshoots it in the opposite direction.

This suggestion can be illustrated by Fig. 2.35 A, B. All of the 'hyperrestoration loops' plotted in a stress–time space consist of three main parts: (a) a rather rapid stress shift caused by an external force (external perturbation) and thus being considered, in relation to a responding embryonic structure, as a passive one; (b) a lag-period, required for a given structure to prepare its active response (this period can well completely or partly coincide with the continuing action of an external force); (c) the active response itself, which proceeds, as a rule, considerably slower than the external perturbation and drives a structure towards the hyperrestoration of the initial stress. As usual, the tensile stresses are taken as positive and the pressure stresses as negative. Frame A illustrates the response either to a sole relaxation of a pre-existing moderate tension (curve 1), or to the relaxation followed by compression (curve 2). The latter situation implies a loss of some amount of water by a sample (as in the case of hydroid polypes), since otherwise the biological samples are practically incompressible. In each of the cases considered the expected response should be that the tension actively increases exceeding its initial value.

Frame B illustrates the idea of a sample's reaction to the external stretching. Expected now is the active tension decrease up to the generation of the internal pressure force (the active branch being extended towards the negative stress values).

It is evident from the very beginning, that the upper and the lower parts of the diagram can be perfectly coupled with each other: the active branch from the loop *A*,1 or *A*,2 while taking place in a certain part of the embryonic body, is to be perceived by another mechanically bound part as an external perturbation (that is, as a passive branch), and vice versa. In any case, the amount of energy spent during the perturbation will be much

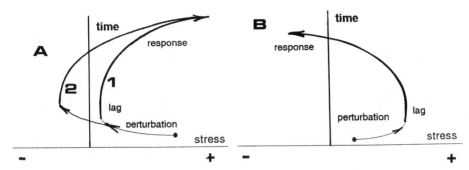

Figure 2.35. The loops of stress hyperrestoration. Horizontal axis: mechanical stress (positive are tensile stresses and negative are compression stresses). Vertical axis: time. For more comments see text.

smaller than that required for the active response.

Similarly to the concept of the active–passive stress feedback, the idea of a hyperrestoration is not new. Either in a hidden or an overt form, it has been exploited in the various fields of applied physiology, such as dealing with the environmental adaptations, sportif training or juvenile growth. In the last respect the ideas of the Russian physiologist, Ilya Arshavsky, are of a special interest. According to his 'energetic rule of motile activity' (Arshavsky, 1982) during the juvenile life period each successive round A of a muscular activity is followed by a storage of a greater amount of energy (available for a next such round B) than that being spent during the round A. By this view, a capacity for a hyperrestoration is a specific and fundamental property of the developing and growing organisms. Obviously, this can take place only in the systems far enough removed from a thermodynamical equilibrium.

Our strategic task for the remaining part of the book will be to explore how the main morphogenetic processes can be derived, in their natural space–time arrangement, from the HRR. We would like to start this long way from anticipating one of the first questions to be asked by an acute reader: what do you mean by an 'initial' stress value (which should be hyperrestored, according to a suggested rule)? Is it a value established in one minute, or in one hour, or in ten hours, before it has been altered by an external perturbation? And how extended will the responding region be? Similarly to the definition of the active – passive stresses, our answer will again be deliberately evasive: it is up to a given dynamic structure to select both the spatial and temporal scale of its response, and it will be just the diversity of the space–temporal scales chosen by the different dynamic structures to endow the development with its richness and plasticity. More than that: as will be argued in this and the next chapter in more details, within a entire developing organism a whole hierarchy of the 'initial stress values' is emerged, each of its members being characterized by a specific life–time (T_{ch}) and embracing an embryonic territory of a specific dimension (L_{ch}). Thus, the main components of each next level T_{ch} -s and L_{ch} -s are created, by our idea, just by the durations and the spatial extensions of HR responses.

2.4.2. MAIN TYPES OF HYPERRESTORATION RESPONSES

Let us review now the HR responses at the different structural levels. We shall start from recalling the responses of the single cells to a stretch (2.1.5). As we know already, they are of two kinds: a sample's contraction (this taking place in the case of non-fixed edges) and the insertion of new portions of a cell membrane, going often via exocytosis (mostly in the case of fixed edges). A contraction response is exemplified by the behaviour of the cytoplasmic strands of *Physarum* (Wohlfarth-Bottermann, 1987), while the insertion response by the active elongation of a stretched neurite (Dennerly *et al.*, 1989). In the latter case the overshoot is obvious. The production of an excessive amount of cell surface materila by a stretched bacterial cell, making possible its division (Harold, 1990), is also a kind of an overshoot reaction, directed to a stress release.

Coming now to the relaxation-promoted reactions at the individual cells' level, one may recall the endocytotic response, which is also directed towards the restoration of the previous stress (now characterized by greater stretching). Obviously, the endo- and exocytosis should be perfectly coupled with each other, each of them compensating the stress shifts caused by the other.

Let us now demonstrate, that all of these reactions have parallels at the level of cell collectives. For doing this we shall review what we know about the reactions of embryonic tissues to the relaxations of pre-existing stresses and to stretching.

The reactions of embryonic epithelia to the relaxations of pre-existing stresses are largely dependent upon the border conditions. Namely:

– if the edges of a layer are immobilized (as being attached to the neighbouring cells or to an underlaid substrate) a typical reaction to the relaxation will be just cell columnarization, associated quite often with formation of the cell's fans because of the domination of the apical contraction over the basal one. So far as this reaction implies the resorbtion of the apical cell membranes, it can be regarded as a kind of a 'cooperative endocytosis'. Perfect cell fans can be easily obtained in abnormal positions if relaxing, by a basal incision either a flat suprablastoporal region or an extensively curved dorsal lip of a gastrula stage amphibian's embryo (Fig. 2.36). In the both cases the cell's reaction is clearly biphasic: firstly (in few seconds after operation) a relaxation of pre-existent tensions is taking place, associated with but a small increase in the cells' length/width (*L/W*) ratios, or a slight relaxatory 'opening' of a dissected lip (frame C, c). This part of the reaction is temperature insensitive. Later on the active, more slow and temperature-dependent reaction branches becomes initiated, which produce clusters of the extensively columnarized and apically contracted cells (frame A, 40'; B). Sometimes they dramatically changing the entire geometry of a layer transforming, for example, an initially convex lip's surface (frame C, b) into a concave one (frame C, d). An overshoot component of the reaction (that is, the production of a stress exceeding its pre-relaxation value) is here quite obvious. Another way for testing this component is in comparing the angles of the passive bendings of tissue pieces dissected before the stress relaxation, immediately after the relaxation, and several dozens of minutes later (regarding the angles' values as rough measures of the relaxed tensile stresses). As seen from Fig. 2.37 A-C) those detected in the latter case exceeded not only the second, but also the first ones. From this we

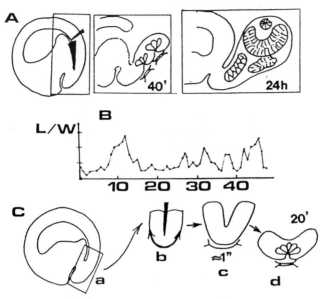

Figure 2.36. Some easily reproduced reactions of embryonic cells to tension relaxations. Incisions are indicated by black edges. A: relaxation of a suprablastoporal part of an early gastula *Xenopus laevis* embryo. 30-40 min after the operation pronounced cell fans are formed on the relaxed surface participating later on in the formation of abnormal neural structures (right frame). B: 1-dimensional map of L/W ratios (vertical axis) of the suprablastoporal cells in 40 min after relaxation. Cells with increased L/W ratio appear as clusters. Horizontal axis: number of cells from the blastoporal level towards anterior C: results of the saggital dissection of an isolated dorsal blastoporal lip (a, b). Firstly a slight, temperature-independent relaxatory opening (c) and then the active temperature-dependent contraction (d) of the previously stretched convex lip's surface is taking place. From Beloussov, 1988, 1990.

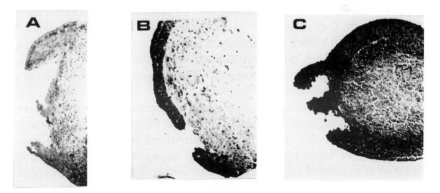

Figure 2.37. A comparison of the bending angles formed by the dissected edges of an ectodermal layer in the intact *X. laevis* early gastrula (A), in few minutes after the relaxation of the tangential tension by a ventral incision of an ectodermal sheet (B) and in about an hour later (C). In all the cases embryos were fixed in 1 min after dissections. The bending angles in C exceed those observed not only in B, but also in A, pointing thus to a *hyper*restoration of the tensions taking place in an hour after their relaxation. From Beloussov and Luchinskaia, 1995a.

conclude, that in few dozens minutes after the overall relaxation of a sample the initial tangential tension has been indeed *hyper*restored.

Another reaction to the relaxation or to a slight compression, typical for the deep cells of the multi-layered epithelia (those non-exposed to the outer surface), consists in the reduction of cell-cell contacts up to a complete rounding of cells and a disaggregation of an epithelia to a mesenchymous cell mass. This takes place not only in the early amphibian embryos but also, in a very extensive manner, in the relaxed neuroepithelia of embryonic brains (experiments by Dr. Saveliev; see Beloussov et al., 1994). Since a reaction is obviously associated with a partial resorbtion of a cell membrane, it is similar to the relaxation-mediated endocytosis. Its peculiarity is in that it is directed towards the hyperincrease of tensions *on the level of individual cells,* rather than an entire cell layer.

– if a layer's edges are free (we are dealing with an isolated piece of embryonic epithelium), the reactions to the tensions relaxation are as follows:

(a) *Smooth sphericalization.* As mentioned before (see Fig. 2.19 and the corresponding comments), a piece of a ventral ectoderm, removed from an embryo at the blastula – early gastrula stage is transformed firstly (within several dozens of minutes) into a compact smooth spherical ball, its surface being lined by so called outer ectodermal, or epiectodermal cells. Whether this phase of the reaction is already associated or not with the increase of the tension on the surface (T_s), depends upon the initial proportions of the piece. As can be shown by simple geometrical calculations based upon a presumption of the preservation of the total sample's volume, only if a diameter/thickness ratio (D/T) of a rectangular tissue piece is less than approximately 10:1, will the initial T_s be exceeded (overshot) by its compact sphericalization, since in this case the surface of a sphere of the same volume will be greater that the outer surface of an initially flat rectangular piece. If, however, a tissue piece is more extended (D/T greater than 10:1), its sphericalization will not lead immediately to T_s being overshot. However, the latter well can be reached now as a result of a subsequent formation of an osmotically swelled cavity within an isolated piece; actually, such a cavity increases a sample's volume by about 1/3 (see Fig. 2.19 C) ; under these conditions the maximal D/T compatible with T_s overshoot, increases up to about 20:1, covering practically all of the real cases. Even in a more straightforward manner the overshoot of the initial tension can be achieved in another simple experiment, employing a capacity of an isolated vegetal hemisphere, isolated from a blastula – early gastrula embryo to a spontaneous spericalization (Beloussov and Snetkova, 1994). As can be shown by elementary calculations, during sphericalization of a solid hemisphere with a naked equatorial surface into a solid ball completely covered by the initial hemisphere's surface the latter will be increased about 1.6-fold. So far as no additional cell material has been inserted into the surface layer we suggest that it is stretched to the same extent. Therefore the initial relaxation of an isolated vegetal hemisphere's surface is followed by its hyperstretching.

(b) *Edge curling.* An ectodermal piece extirpated from an the late gastrula - neurula stages embryos restores (and then overshoots) the relaxed tensions by curling itself around its edges. This is accompanied by a columnarization and a subsequent immigration of a substantial number of cells and also by the formation of a pressurized cavity inside. Consequently this complicated response includes all of the above mentioned elementary responses. A real overshoot of the initial tension can be indicated in these cases both by the dissection tests and by measuring the edge angles (Beloussov *et al.*, 1994).

The reactions to a sample's *compression* can be illustrated, as told before, by the pulsating cells of the hydroid polypes. The above desribed flow of water out of the naturally or artificially compressed cells as well as the extensive self-stretching of the cells at the retraction phase of the pulsatorial growth clearly indicate the cell's tendency to restore the pre–compression stresses. It should be meanwhile the matter of further investigations, whether these reactions are associated with the real overshoots.

Reactions to stretching are also largely dependent upon the border conditions. What is now decisive is whether a stretched sample is capable of restoring, and even hyperrestoring its initial stress value by a mere contraction, or, on the contrary, the stretching force is so great and/or acts for such a long period that this becomes impossible. Let us consider the following situations:

(1) A stretching force is rather small, or it acts within a not so large time period, so that a sample is able, after some time delay at least, to (hyper)restore its initial stress by contracting both its ends. This will be a stretch–contraction response (or relay), firstly postulated by Odell's *et al* (1981). In a very clear way it is exhibited by the cytoplasmic strands of *Physarum* (see p.91), although a number of other processes also seem to go according the same pattern. For example, an ubiquitous response of the cultured cells onto the substrate stretching is their orientation perpendicularly to a stretching force (see, for example, Dartsch and Hammerle, 1986; Levina *et al.*, 1996). This may be interpreted as a direct result of the cells' contraction just along the directions of their previous stretching. Several examples of similar responses are shown by embryonic tissues. Condic *et al.* (1991) consider a contraction of the pre–stretched cells of the imaginal disc in insect larvae as the main tool for this rudiment's morphogenesis. Our model data (Beloussov and Lakirev, 1988) makes it highly probable that the post-stretching contraction is involved in the morphogenesis of the axial rudiments in amphibian embryos. It is worth mentioning that the Belintzev's model, although at first glance it looks like the opposite of that of Odell *et al.*, in no way excludes a post-stretching tangential contraction participating in the contact cell polarization, at the conditions $p_i > \langle p \rangle$ (see pp. 129-131). And last, but not least, the above discussed quasirelaxational character of epithelial folding should be also considered as a kind of a contraction response to a previous stretching of the fan's cells, partly shadowed here by the fact that the contraction now leads to a change of geometry, rather than to a simple shortening of a layer.

(2) A strictly localized stretching (pulling) force is applied to one of the sample's poles (*A*) while the reaction force at the opposite pole *B* is smoothed, as being dispersed among many material bonds. In such a way a tension gradient (see p. 87, Fig. 2.4 D and the corresponding comments) is established, going down in *AB* direction. The expected HR responses should be directed towards smoothing the gradient; this can be achieved, for example, by the active shifts of cells from *B* to *A*. This appeared to be the case: already in few minutes after stretching a piece of the ventral ectoderm of early gastrula *Xenopus* embryos its cells started to move actively and in a cooperated manner, upwards the gradient, that is towards the source of a pulling force (Beloussov and Louchinskaia, in preparation). Such a reaction which may be called 'tensotaxis' can be responsible for a

large set of directed cell movements during embryonic morphogenesis (see Chapter 3 for
further discussion).

(3) Both ends of a stretched sample are firmly fixed, so that it is incapable of restoring
its initial stress value by a contraction. This is one of the most important morphogenetic
situations making available two different, although mutually related ways for
(hyper)restoring the stretching stress. The first one lies in producing (engulfing,
synthesizing) a new material which will be inserted into the tensed structures, while the
second one lies in rearranging an already existing material. The first way is exhibited in
its clearest form in plants' growth, and is described by the already discussed Lockhart's
law: the stresses are (hyper)relaxed as a result of the insertion of the cell walls subunits,
this insertion providing both the linear growth and morphogenesis. In animal species a
similar way is provided by stretch-induced cell divisions (and, hence, post-mitotic cell
growth), their existence being demonstrated in several cases (Opas, 1994; Singhvi et al.,
1994). Obviously, the post-mitotical cell growth is directed just to the diminishment of the
imposed stress. In most cases, meanwhile, this relatively slow way is exchanged, or, at
least, supplemented, by a direct redistribution of an already existing material in such a way
that a sample is shrunk in the direction perpendicular to stretching. If assuming that this
reaction is performed by single cells having no neighbours for a mutual repacking
(suggest, for example, that we stretch a cell line with one cell in width), it is only the
intercellular material to become redistributed. Under these conditions, the cells should
narrow themselves perpendicularly to the stretching direction and become elongated
parallel to it. Meanwhile, as soon as the stretched rudiment will have more than one cell in
width, a more economical way, consisting of mutually repacking (intercalating) its cells in
the direction perpendicular to the stretching is opened up. This is what may be defined
as transverse cell intercalation.

We were able to trace the both latter reactions in the above mentioned experiments by
stretching the explants of the early gastrula animal ectoderm from *Xenopus laevis* embryos
(Beloussov and Luchinskaia, 1995). Simultaneously with the above described 'tensotaxis'
movements (and as a part of them) the first (intracellular) phase of the post–stretching
reaction could be traced. It was exemplified by the appearance, on the surface of a 2 –
2.5-fold stretched explant, of the cells elongated in the stretching direction to a much
greater extent than an entire explant is (Fig. 2.38, cf. B and D). Such an excessive cell's
stretching points to the presence of the active intercellular extending forces, directed
along the imposed stretching but exceeding the latter. Similarly, at the same time period
the cells start to produce prolonged protrusions, which are also elongated in the stretching
direction, and continue to do this well after the stretching is over. The formation of these
protrusions indicates the overproduction of the cell's membrane. The next phase of the
reaction, that is, transverse cell intercalation, comes into play a bit later, in about an hour
after the cessation of stretching. At this time a sample continues to become schrinked
perpendicularly to the stretching direction, almost all of its cells restoring their initial
isodiametrical shapes (Fig. 2.38, cf F and B). As shown by videofilming, this period is
associated with both transverse and vertical cell intercalation, going on in the different
locations throughout an entire surface of the explant. At the same time, the polar explant's
regions form inflations, in some cases perforated (Fig. 2.38 E, G).

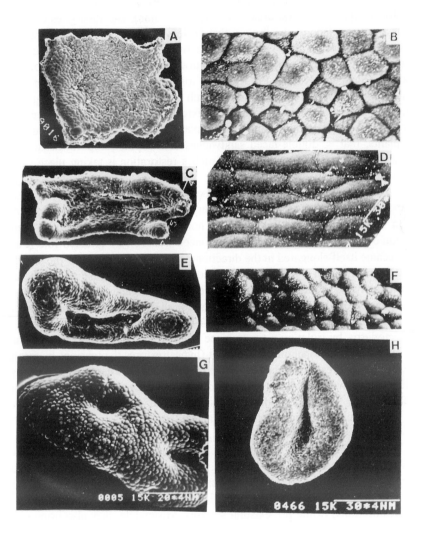

Figure 2.38. Experiments on stretching the explants of the animal ectoderm extirpated from *Xenopus laevis* early gastrula embryos. A,B: before the start of a stretching, C,D: immediately after cessation of about 30 min duration, 2.5-fold stretch, E,F,G: 5 h later, without additional stretching. H: a sample which was brought back into the relaxed state immediately after cessation of stretching and kept in these conditions for about 5 hrs. A,C,E,F,G,H: general views of explants from the top (i.e. from the substrate-opposed, epiectodermal sides). B, D, F: some representative epiectodermal areas of the samples, shown at left. Note that the cells in F, after being extensively elongated (D), now return to their isodiametrical shapes, similar to those in B. For more details see Beloussov and Louchinskaia, 1995.

The above described stretching experiments indicate that the active cell responses to a stretch, although starting during the stretching procedure itself, reach their full expression mostly after the cessation of external stretching. That relates, most of all, to the intercalatory response. To become initiated it requires a certain minimal duration of stretching (for a suprablastoporal area this is about 30 min), and then goes on as if 'by inertia'. It is a proper place to recall now, that no less than about 30 minutes presence of an outer ectodermal layer at the very start of gastrulation is both necessary and sufficient for the initiation of a convergent cell intercalation within the deep tissues of a dorsal blastoporal lip (Shih and Keller, 1992). One may suggest that the role of an outer ectodermal layer is just to bear tensions which are required for switching on the intercalation movements. This layer is not only extensively tensed in normal embryos, but also restores tensions quite soon after their relaxation: by our observations, in so called Keller's explants, that is, double-faced sandwiches of a suprablastoporal tissue prepared from the early gastrula amphibian embryos, such a restoration is taking place very soon after the isolation of an explant.

Besides the post-stretching reactions themselves, we have looked how a previously stretched explant will response to its secondary relaxation, oriented in the stretching direction. In these experiments, within 4 – 6 hours after the cessation of a secondary relaxation, the explant either produced a pronounced transversely oriented fold (Fig. 2.38 H), and/or became itself elongated in the direction perpendicular to that of the stretching – relaxation. Obviously, such a reaction is directed towards the increase of the secondarily relaxed tensions, that is to the restoration of the previously established stress regime.

The responses of the ectodermal explants to all of these manipulations can be summarized by the following diagram (Fig. 2.39). Firstly, we relax the tensions in the embryonic ectoderm by isolating a tissue piece from an intact embryo. While being kept isolately without any additional interventions, it (hyper)restores the tensions by means of a sphericalization and a creation of the pressurized cavity inside, as descibed previously (Fig. 2.39 A). If however we prevent the sphericalization and a cavity formation by stretching the piece artificially, we get the different kinds of the above described intercalatory reactions, all of them, producing, in the mechanical terms, a certain longitudinal 'active pressure' within a piece (Fig. 2.39 B); at the same time, the polar regions of the tissue piece will look as being actively contracted. At last, if we intervene in this process by secondarily relaxing a sample, it will create a pronounced fold, oriented, as mentioned before, perpendicularly to the stretching - release directon (Fig. 2.39 C). As seen from the Fig. 2.39, all of these reactions can be qualified as at least the restorations (and in many cases hyperrestorations) of the pre-existing stresses. Meanwhile, they also bring us to the following, more specific, conclusions:

– an embryonic tissue reacts each time to the *last* external stress perturbation, 'forgetting' those taking place before it. In order to be remembered, the last stress regime imposed onto *Xenopus* embryonic tissues should have a duration no less than 30 – 60 min. The same conclusion could be also driven from our earlier (Beloussov et al., 1988) experiments on stretching the suprablastoporal tissues explants of *Xenopus laevis* embryos perpendicularly to the direction of its normal (antero-posterior) stretch. Just 30 min stretching appeared to be a minimal time for producing a perfect set of the correspondingly reoriented axial organs. Further reduction of a stretch duration led

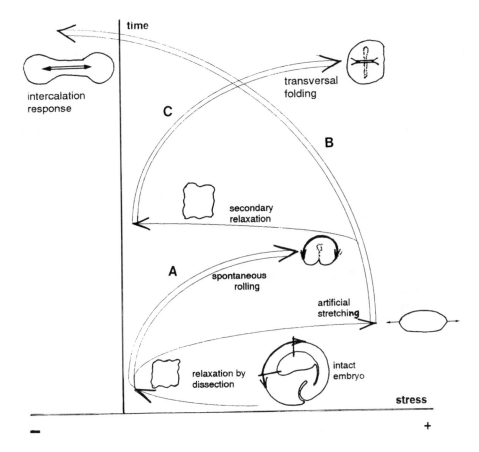

Figure 2.39. A succession of the reactions of an explant of the animal ectoderm from *X.laevis* early gastrula embryos presented as a chain of the hyperrestorations of the previously established stresses, each time disturbed by a next intervention. Double-lined curves indicate active hyperrestorative responses and one-lined curves the experimental modulations of stresses. For more comments see text.

progressively to ever less structurized rudiments and under zero stretch the explants immobilized on the latex film, have been completely dissociated into single cells.

 – a modulation of a stress regime, caused by an external force is perceived by a tissue piece in quite a different way from a stress which is actively generated by the tissue itself, even if both are directed towards the same side in a stress/strain phase space. The main point is, that any externally imposed stress shift $\Delta\sigma$ is followed by an oppositely directed hyperrestorative response (as a kind of a 'protest' of a forced tissue), even if $\Delta\sigma$ is of the same direction as the stress shifts produced *actively* by the same sample and at the same time period. Coming back to our diagram we can see, that the morphogenetic response of a pre-stretched tissue piece to the externally imposed tensions'relaxation (Fig. 2.39 C) is quite different from that produced by its inherent (intercalatory–mediated) relaxation tendency. In the other words, what is created by an *internal* activity, can in no way be substituted by *outside* influences, even if they seem to be directed towards the same goal (the reader can easily invent a lot of psychological and sociological examples of this kind). In the long run, meanwhile, such external violence appears to be for the good: as we hope to demonstrate in Chapter 3, an entire morphogenesis can be regarded as a succession of 'protests' of embryonic tissues to the stress shifts caused by external influences. Directed towards the never achievable previous stress patterns, they nevertheless drive the development forward. Such is one of the paradoxes of the Bergsonian 'active memory'!

It is of interest that a powerful cell intercalation response can be obtained in the same stage amphibian embryos as a kind of an indirect (secondary) response to their artificial relaxations. This becomes possible because the relaxations are followed by a hyperrestoration of tensions, mostly due to the tissue's curling around the wound edges. As a result of such a retension, cell intercalation will be initiated, actively elongating the previously relaxed part(s) just in the curling's direction. In the other words, instead of being stretched artificially (as in previously described experiment) the embryonic tissues become now self-stretched. Because of high developmental potencies of the early gastrula tissues, the secondarily stretched areas, whatever be their locations, often produce axial structures (Fig. 2.40).

As indicated by some unfortunately up to now scattered data, even the samples taxonomically far removed from amphibian embryos reveal nevertheless fairly similar stress–dependent patterns. Let us mention just three examples, very much differing from each other:

1. In the growing buds of hydroid polypes, the artificial stretching of an ectodermal layer (by increasing a hydrostatic pressure within the gastral cavity) leads, within no more than a few minutes, to the increase in the number and the volume of the spherical vacuoles within the cells (Kazakova *et al.*, 1997). Obviously, such a reaction is directed to the increase of the internal pressure stresses and can be regarded thus as mechanically similar to the cell intercalation. Some additional examples of the stretch-induced morphogenetic reactions in hydroid polypes will be discussed in Chapter 3.

2. In the aggregates of a chicken embryonic heart's mesoderm stretched by centrifugation, an unusual process of so called cell emigration, that is the penetration of the deep cells into an external layer, has been described (Steinberg, 1978). Similarly to

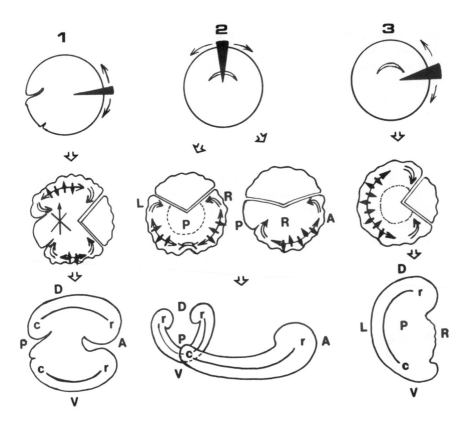

Figure 2.40. Producing a complicated morphomechanical reaction: artificial relaxation – autonomous retension – cell intercalation. Tensile stresses on the surface of *Xenopus laevis* early gastula embryos have been relaxed by inserting additional tissue pieces in various locations. Upper row: (1) insertion to the ventral side (opposite to blastopore), (2): insertion to the dorso-medial area, (3): insertion to the lateral side. Second row: scheme of cell movements directed towards hyperrestoration of the relaxed stresses. Curling of tissue edges is shown by small curved arrows, cell intercalation by short oppositely directed arrows and intercalation-mediated extensions by prolonged double-head arrows. Third row: schematical representation of the results. Additional axial structures formed in different abnormal locations are shown by c-r lines where c is a caudal and r is a rostral pole. As depending upon the site of a relaxation, the axial structures may become duplicated (left frame), triplicated (middle frame) or shifted to the lateral (right frame). A: anterior, P: posterior, D: dorsal, V: ventral, L: left, R: right sides of the host embryos. From Beloussov and Louchinskaia, 1995.

the previous example, this process is directed, in its own way, to the diminishing of the artificially imposed tensions. Unfortunately, the description of these beautiful experiments is not enough complete for deciding unambiguously, whether the cell emigration was intense to the extent permitting to produce a real overshoot response (generation of a cell–cell pressure), or whether it only reduces the imposed tension without transforming them into internal pressure.

3. As concerning the stretch therapy of the bone malformations in humans, effectively performed, for example, in Professor Ilizarov's clinics in the Siberian town of Tyumen, it is the post-stretching relaxation which is used as the most suitable criterion of a successful cure. The best results always correlate with an extensive (and directly measurable) active relaxation of the imposed stretching stresses by the tissue itself (Ilizarov, 1984). What we have here, definitely fits the reaction loop shown in Fig. 2.35 B.

2.4.3. A MULTILEVEL HIERARCHY OF HR RESPONSES: ALTERNATIONS IN SYMMETRY TRENDS AND GENERATION OF FRACTALS

Consider a typical HR reaction to the relaxation of pre-existent stresses. While observed at a certain macroscopic level it is associated, as a rule with a reduction of a symmetry order because of the formation of numerous folds, ruffles, villae, domains or fans of columnarized cells, etc.(see Fig. 2.36 A, Fig. 3.7 C, D). Each separate fold, domain or a fan bears meanwhile within itself a perfect radial or, at least, mirror symmetry. Meanwhile, the individual cells comprising domains or fans are again largely asymmetric because of their apical contraction, oblique shapes and so on. We can conclude therefore that each next level is characterized by the opposite trend in changing its symmetry order.

On the contrary, a stretching of a tissue piece largely smoothes out (during the active HR reaction phase) its contours, thus increasing a symmetry order of an entire sample. As a postponed reaction, meanwhile, at least its translational symmetry order can be decreased, because of the achieved bipolarity (Fig. 2.38 E - G). At even lower level we can trace largely isodiametric cells with rounded outlines which increase again their symmetry order. So, the same alteration of symmetry changes takes place as well.

The situation can be also analysed in the following manner. Suggest that a HR reaction of an entire embryo to a stress relaxation is mediated by the formation of some macroscopic folds, as actually observed in relaxed blastulae (see Fig. 3.7 C, D). It is easy to calculate that when a newly arised fold changes its shape from a roughly hemispherical to a roughly cylindrical while continuing to enclose the same volume (same amount of a cell material) its surface area will be increased and thus in the first instant stretched. The stretching will be even greater if a fold takes a flattened (blade-like) shape, as often happens. Correspondingly, within each individual fold's scale a HR reaction to stretching, that is, a production of an excess of a new surface material can be now expected. As a result, the smaller (next order) folds will be produced, and so on. In such a way we can expect an entire hierarchy of the progressively diminished folds, that is a kind of a fractal structure which can go down well below the optical level. Some examples of such a situation, as well as a discussion about its possible evolutionary consequences will be presented in Chapter 3.

2. 5. Standard morphomechanical situations

In this section we would like to generalize what has been said before about the stress responses by outlining the typical, or standard, *morphomechanical situations* which are numerously reproduced within morphogenesis and are naturally linked to each other. To describe an entire morphogenesis as a function of such a linkage will be the task of the next part; what we have to do now is just to describe each of them separately. What is common for all of them is that they take place only in the mechanically continuous and stressed contours.

2.5.1. CONTRACTION–EXTENSION POSITIVE FEEDBACK (C–E FEEDBACK)

C–E feedbacks in a tensed cell net: instability of equitensed (120°) nodules.
Let us take a third order equitensed nodule (Fig. 2.41 A) and exert a small *active* stress perturbation along some of its fibers. This can be done in one of the following ways:

(1) one of the fibers (say, *a*) is actively contracted, thus stretching two others (*b* and *c*) – Fig. 2.41 B;

(2) two fibers (*b* and *c*) are more or less equally contracted, stretching the third one (*a*) – Fig. 2.41 C;

(3) one of the fibers (*a*) is actively extended, relaxing two others (*b* and *c*) – – Fig. 2.41 D.

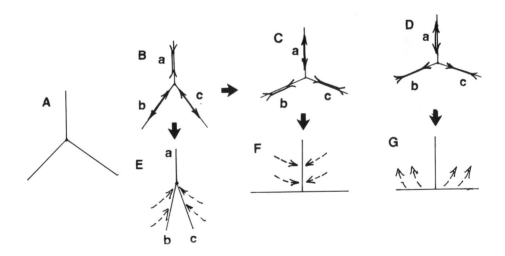

Figure 2.41. Possible types of the contraction-extension feedbacks within a tensed net (say, of cell walls). Dotted arrows indicate the insertion/resorbtion of cell material.

Situation (1) will lead, by HR idea, either to the active contraction of the fibers b and c, that is to the situation (2) (Fig. 2.41 C) or (in the case of a strong mechanical resistance preventing the contraction) to the insertion of some new material within the same fibers (Fig. 2.41 E). In any case, the initial tensions' equality will be perturbed, so that the angles between the fibers will deviate from 120°. In the other words, a certain combinations of two fibers (a–b, a–c or b–c) will become more straight than before. In such a way the delamination-like processes can be modelled. It is also worth mentioning, that each one of the transitions B → E and C → F(Fig. 2.41) reproduces the above described phenomenon of tensotaxis, that is, the active cell's movement along the direction of an external pulling force. In the first case (B → E) a cell will move upwards (from b,c which imitate a cell's leading edge, towards a) while in the second case (C → F) it will move in a reverse direction, the external pulling force being imitated by the contraction of the fibers b,c.

C–E feedbacks in a closed tensed contour: establishment and a dynamical maintenance of a polarization – Fig. 2.42 A.
Let us suggest that a region a of a closed, and at first moderately stretched, contour is contracted, either actively or passively. In any case, it will stretch the connected parts of the contour. If the latter tend to hyperrestore their initial stress values and are incapable of overriding the contraction force coming from a, they will do so by inserting some new material, thus relaxing a. According to the HR idea, this will promote further the contraction of the piece a, and so on. Such a feedback loop should require the inflow of some material towards b and will become arrested with the exhaustion of the flow. In the case of a direct coupling between the contracted (material–resorbing) region a and an extended (material–inserting) region b the situation will be self-perpetuating, the material flow being directed from a to b. The locations of both zones will be stably retained. That means that a sample first acquires and then retains its dynamical polarity. This seems to be the most general morphomechanical scheme of dynamical polarization and polarity maintenance working on quite different structural levels, from a single cell up to a multicellular organism.

C–E feedback within a more or less planar (non-increasing its curvature) sample, for example within a part of a cell sheet (a uniplanar C-E feedback) –Fig. 2.42 B.
Let us consider a compressible cell sheet which can be effectively stretched by an external force, much less effectively stretched by an internal pressure and which cannot be shrunk at all because of its semi-fixed edges. Such a situation is typical most of all for embryonic tissues. According to the above, a stretching of such a sheet from the outside should switch on cell intercalation, oriented perpendicularly to the stretching force (Fig. 2.42 B_1) and thus making the sheet thinner. For reasons of symmetry the thinning will be the greatest in the central region of a sheet (equally removed from its edges), the compression wave generated by intercalation moving towards the sheet's edges. Consequently, cell material will be concentrated and at least relaxed (if not compressed) near the opposite poles (Fig. 2.42 B_2). According to the HR idea, this will switch on the active contraction of the polar regions, possibly associated with the immigration and/or cooperative involution of at least some cells (as shown by our experiments, see

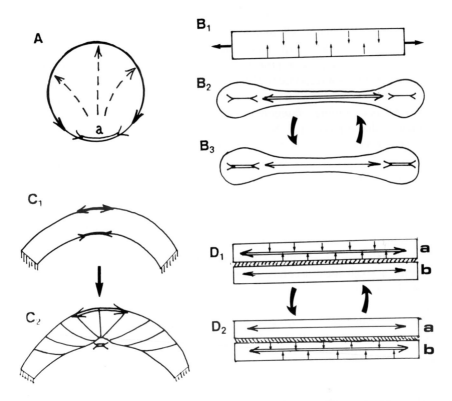

Figure 2.42. Contraction-extension feedback in the closed contours and cell sheets. For comments see text.

Fig. 2.38 G). Any one of these polar regions' activities will stretch further the central part, promoting a continuing intercalation, the latter again stimulating polar contraction, etc. (Fig. 2.42 B₃). Such a positive feedback will work until the material capable of intercalation and/or immigration will be exhausted. At this time the transition between the extension and the contraction zones may be very sharp (cf. Fig. 2.6 C). It is noteworthy, that at the moment of exhaustion (and, hence, of the intercalation arrest) we come exactly to the 'Belintzev situation'(interaction of the tangentially contracting cells with a stretched part of a cell layer without a possibility for further cells' repacking) with all of its self-organizing properties. In this way an initially homogeneous piece of a sheet with symmetrical border and initial conditions will be transformed into a bipolar structure with its poles inflated and probably bearing invagination holes and the central part elongated and narrowed. That fits well with the above described results of the stretching experiments (Fig. 2.38) and corresponds to the most fundamental and ubiquitous morphogenetic events, taking place, for example, during gastrulation and neurulation of vertebrate embryos (see for more details Chapter 3).

It seems probable that a similar C–E feedback is also participating in the stress–mediated self–organizing phenomenae in cell cultures, as described by Harris and coworkers. For example, the cell files shown in Fig. 2.16 C at the right may correspond to the extended part while a cell aggregation (the same frame at left) to the contracted part. According to such an interpretation, the formation of the stretched cell files cannot be regarded as a purely passive process of cell traction by the stretched substrate: the active cell extension, as a response to the applied stress, should be also in play here.

C–E feedback in a folding (changing its curvature) epithelial sheet – Fig. 2.42 C.
Let us suggest that a piece of a sheet with its edges fixed is slightly bent by an external force. In the general case its concave surface will be relaxed or even compressed while the convex surface will be passively extended (stretched) – Fig. 2.42 C_1. According to the HR idea this will promote a subsequent active contraction of the concave side and an active extension of the convex side. This may well be associated with the flow of material (either intracellular components or entire cells) from the concave to the convex side. As a result, along with an overall increase of a sheet's curvature, that of the convex side will differ from the curvature of the concave side in the way shown in Fig. 2.42 C_2. Namely, on the macroscopic scale the convex side's curvature will exceed that of the concave side, while on the latter a small local secondary fold, flanked by more flattened regions, will be produced. Later on (Chapter 3) we will present some real examples of this wide spread configuration.

By just the same reasons (a side slightly concaved by an external force tends to become more concave while that slightly convexed tends to become even more convex) any symmetrical evagination will be unstable tending to be bent in a constant direction. That means, in particular, that a slightly asymmetrical evagination formed on the surface of a large 'maternal' rudiment has an inherent tendency to be pressed anyhow towards this rudiment's surface. Same tendency is amplified by the unequal contraction of the oppositely located basal axiles of an invagination: the axile which is initially even slightly narrower than the opposite one will continue to diminish its angular value more actively than the opposite one. This may explain a wide presentation of tightly compressed rudiments, from plant leafs to amniotic folds of animal's embryos. On the contrary, each time when a 'daughter' rudiment is deviated from a maternal one a specific additional agent is required (see Chapter 3, pp. 206-208)

2.5.2. EXTENSION–EXTENSION POSITIVE FEEDBACK (E–E FEEDBACK)

Generally, such a situation takes place in two or more tightly bound morphological entities (from cells to multicellular sheets), responding to the external stretching by inserting (intercalating) new material in the direction perpendicular to stretching, and thus relaxing the latter.

For animal embryos a typical scenario is as follows. Suppose that a sheet *a* is stretched passively (by an external force). This will stimulate its active (intercalation-mediated) extension. By doing this it will stretch the attached layer *b*, initiating there the same active reaction, and vice versa (Fig. 2.42 D). In this way a mutual aid between layers is established. This feedback will produce much more effective and coordinated extension than that occuring in each layer separately. In vertebrate embryos such a feedback may coordinate the epiboly of an outer layer with the involution of the inner one: it was, probably, one of the most effective evolutionary inventions of this taxonomic group.

At the single cell level a similar situation can take place both during the contact cell polarization and, also, in plant tissues. As has already been mentioned, the stretch dependence of the cell wall material's insertion is the main rule of plants' growth. Now we may add to this that insofar as the adjacent plant cells almost inevitably are growing at somewhat different rates, being at the same time tightly bound with each other, they should necessarily stretch each other, adding this mutual stretching to an internally borne one. The consequence will be a mutual promotion of the membrane's insertion and, hence, of the coordinated cell's growth.

2.5.3. CONTRACTION–CONTRACTION POSITIVE FEEDBACK (C–C FEEDBACK)

In this case two topologically different situations can be distinguished:

(1) C–C feedbacks which take place just along the contraction plane. This is a frequently discussed relay situation, fitting with both the Odell *et al.* and Belintzev's models: contraction of the cell A extends one or few of the adjacent B cells, which have enough power for (hyper)restoring the initial stress value by their subsequent active contraction (rather than by the redistribution of a material, as in the case of C–E feedback), etc.. More precisely, we can define this situation as C–E–C feedback.

(2) C–C feedbacks going on perpendicularly to the contraction plane. Suppose that we have two superimposed mechanically bound cell layers a and b (like those shown in Fig. 2.42 D) one of them (a) one of them being now actively contracted. If the mechanical bonds between both layers are strong enough, a superimposed part of a b layer will be compressed, relaxed or at least not extended. It is worth mentioning that, because of b–a anchoring, if b cells are proliferating this layer will become effectively relaxed. In any case, the conditions for the active contraction of b layer are provided. In the case of the apico-basal asymmetry of the b layer's cells this contraction will lead to its folding. The following examples of such a situation can be given: (1) a mutual enhancement of the folding of the axial mesoderm and the neural plate along their dorsal midline; (2) similar folding's enhancement in the initially removed cell layers after they have established firm contacts with each other. Such events take place, for example, in a foregut endoderm and the oral ectoderm of many Deuterostomia embryos, as well as in a lens placode and the adjacent part of an eye vesicle. The experimentally proved possibility of producing an involuting placode in an unusual location by preventing mechanically the free expansion (stretching) of the ectodermal cells (Steding, 1967) supports this idea.

2.6. Summary of the chapter

In this Chapter we have explored, starting from the supramolecular scale and going in an ascending order, the dynamic structures involved in the developmental processes. We have shown that it is possible, indeed, to consider the whole set of these structures as a coherent spatio–temporal hierarchy. Our account was very much mechanically oriented, one of our leading ideas being that the *macroscopic* forces and deformations are largely involved in the morphogenetical feedbacks. On the other hand, we have considerably deviated from classical 'non-biological' mechanics by introducing the concepts of passive and active mechanical stresses. We have presented numerous examples of the mechanosensitivity of stress generating devices and traced the ways for the mechanical stresses to be stored in and transmitted throughout embryonic tissues.

At the single cell level, we have described the dynamic structures involved in cell polarization, cell movement and cytotomy, mainly looking for what is in common between all of these processes. Then we have passed to the cell collectives, starting our review from freely migrating cells of slime moulds and moving, via epithelioid cell masses towards the real epithelia. We have treated both cell immigration and cell intercalation from a common point of view, arguing that they differ from each other only by their global, and not the local description. Later on we approached the often neglected delamination (cells alignment) processes which are a powerful tool for early embryonic patterning. Coming to the epithelial morphogenesis, we have paid special attention to the formation of columnarized cell domains and to so called 'cell fans', pointing out that they are, on the one hand, self-organizing dynamical structures and, on the other hand, the stores of morphogenetic information for a finite subsequent period of development. This period is associated with a 'quasi-relaxation' of the mechanical stresses, embedded in these structures: this is an energy consuming movement along a relaxatory pathway. We have presented some arguments showing the biological advantages of a 'quasi-relaxational strategy' as participating in such various processes as the epithelial folding, translational cell shifts and delamination. We have also described a remarkable process of the pulsatorial growth and morphogenesis as an evolutionary ancient example of the translational shifts and folding of epithelia.

Then we have returned to the concept of stress-mediated morphogenetic feedbacks and formulate a phenomenological rule of stress hyperrestoration. We have classified the stress responses of embryonic tissues by formulating several standard morphomechanical situations which embrace most of the known morphogenetic processes. In this way we have prepared ourselves for passing, in the last Chapter of the book, from the single dynamic structures taken isolated from each other to the actually observed successions of morphogenetic events, that is to a real dynamic architecture of development.

CHAPTER 3
THE DEVELOPMENTAL SUCCESSIONS

3.1. Embracing laws for developmental successions: formulating a task.

The previous chapter was mostly analytical: its main aim was to outline, in ascending order, the different levels' processes involved in the development without paying much attention to how they are coupled with each other in space and time. Now we would like to move towards some kind of a synthesis: namely, we wish to understand why the developmental events are arranged in a definite spatio-temporal order. How can we do this?

The human mind has developed two alternative ways for resolving tasks of such a kind. The first of them can be qualified as an additive or, if employing a neo-Kantian philosophical terminology (invented by the German philosopher Wilhelm Windelband), ideogaphic approach. For interpreting, by using such an approach, a more or less prolonged chain of events (including a succession of developmental stages) one should independently discover and then simply enumerate the 'causes' of each successive stage to stage transition without aiming to find something in common between the different causes and, moreover, between the different developmental successions. In other words, we should learn any next succession, so to say, by heart.

The alternative way is defined, in Windelband's terms, as a nomothetic. Now, instead of enumerating the heterogeneous events one after one, a researcher, after some period of initial exploration, should formulate a unifying, embracing law (given, best of all, in mathematical terms) from which an entire succession can be derived with a certain degree of precision. By this approach some of the observed details should be estimated as non-important and thus neglected, while a number of others will be designated as unknowns $X, Y..$, their intrinsic nature being, up to some more or less remote future, of no interest for us. Now our main pursuit will be not in discovering as many as possible different variables (this being the case for the additive, or ideographic approach), but, according to the main ideas of symmetry theory (recall Chapter 1), in finding invariance within the variables, that is in minimizing (instead of maximizing) our knowledge.

It is well accepted that it is the second way which, in spite of its seeming arbitrariness and the much greater role played in this case by the explorer's individuality, was, during at least the last three centuries, the main method of the natural sciences. It was just this way which endowed, in Francis Bacon's words, human's knowledge with its real power. Accordingly, any manifestations of an additive (ideographic) approach in the physicalistic sciences have been considered in the best case as a compelling and preliminary stage, to be later on displaced by a nomothetical theory.

Quite another, meanwhile, was the situation in the humanitarian and social sciences: several attempts at constructing some kind of nomothetic laws for, say, the events of human history or, the more, for the masterpieces of art ('how to make poems' rules, fashionable between the modernists at the eve of our century) yielded almost nothing except a far-going impoverishment of a studied object.

Now, what about the life sciences? The situation in biology may be qualified as more or less intermediate between both extremes. Meanwhile, the psychological attitude (even if deeply hidden) towards an ideographic approach, at the expense of a nomothetic one, is in biology unproportionally high. To illustrate this statement, it is enough to compare the scientific destiny of two outstanding concepts of last century embryology: that of Driesch law ('the fate of an element is a function of its position', see Chapter 1 for the detailed analysis) and of Spemann's discovery of a 'primary organizer' in amphibian's embryos. The latter (irrespectively of a certain Spemann's personal attitude towards holistic approaches) is clearly ideographic, or additive: what was actually discovered, was a 'cause' of no more than a single developmental transition, since each successive 'secondary', 'tertiary', etc. organizer could not be expected, *ex definitio*, to have much in common with the primary one (or, if some of them appeared to be identical, how can they induce different structures? And if, inspite of their idenity, they really do this, what is the reason to treat them as the specific 'causes' of these structures formation?). What was actually discovered by the German scientist, was just a single 'developmental point' within an innumerable multiplicity of others. And nevertheless Spemann's discovery was widely welcomed and crowned with a Nobel Prize, one among very few to be awarded to developmental biologists.

On the other hand, Driesch law should be considered, by all the estimations, as a unique example of a really nomothetic embryological generalizations, with a very wide range of applicability: it is, on the one hand, almost universal (both in the taxonomical and in the ontogenetical aspect), and, on the other, leaves to an explorer a great deal of freedom in establishing the levels and the 'mechanisms' of its action: what is quite typical for the nomothetic constructions, the most important components of the Driesch's law (including, first of all, the idea of a 'whole') are just slightly outlined rather than explicitly described. Meanwhile, exactly these aspects became the matters of many misunderstandings, condemnations, and a prolonged oblivion, followed by a strange reincarnation in the framework of the 'positional information' concept, rather restricted and to a great extent ideographic version of the Driesch's ideas (the main task for many authors is to discover a specific chemical nature of the factors of positional infomation).

It would certainly be senseless to reproach anybody personally for the events moving along such a route: this is the way things are going on, and nobody can neglect that the most flourishing research trends in the present day developmental biology and the related branches definitely follow an ideographic Spemann's way rather than a nomothetical one exemplified by the Driesch law. Such is, for example, modern developmental genetics, regarding its main goal in identifying the genes expressed at the successive stages of development. Within this framework, the question of why a certain succession of the developmental stages is really taking place is either completely neglected or is regarded to be exhaustively derivable from the properties of the genes themselves.

Is such a situation satisfactory and having good perspectives for future? Or it may be compared with a long-termed flourishing of an ideographic trend in the medieval science, followed later on by the nomothetic breakthrought (what we mean here is a replacement of the Ptolemean epicycles by Kepler's laws and then by Newtonian mechanics)? One should not forget that only for the explanation of fully deterministic systems, obeying 'one cause – one effect' rule, a monopolistic usage of the ideographic approach can be more or less justified. But we have presented many arguments (see Chapter 1) showing that the developing systems are quite far from a deterministic image. So why, under these circumstances, should we avoid even a small chance to bring into biology but a minor particle of scientific beauty and mental satisfaction, which accompanies any step of a nomothetic advance? However, we must be very cautious by following this way.

First of all we should be aware from the very beginning that even in the best case a nomothetic approach will not engulf an entire biology. At any time there should be in this science a substantial amount of what we have to accept as given and 'learnt by heart', rather than derived from any universal rule. This is because a historical component, associated with the fixation of something that accidentally happens and hence is unpredictable, will inevitably persist in any aspect of biology: above all, it comes from the very phylosophy of self-organization. A most 'nomothetic' thing which we may do in these cases is to outline nothing more than the field of *possibilities*, or *potencies* of the system under study, without claiming what exactly it will do in reality. This seems to be especially true for the evolutionary theory.

Secondly, we have to accept that at the present time, at least, we are unable to formulate the nomothetic laws in developmental biology in strict mathematical terms. We even do not know exactly whether they should take the form of classical differential equations, or finite element models, or something else. These questions are far from being purely formal: they are associated, first of all, with the fundamental and still unsolved problem of a developmental memory. It would probably be wisest of all to restrict ourselves at the moment to some purely verbal constructions prepared meanwhile in a way keeping a possibility for more precise formulations.

Now we come to our main assumption. We will accept as a working hypothesis that these formulations will be *morphomechanical* in their essence, that is, implying macromorphology and the mechanical stresses among their basic components, and using the idea of stress hyperrestoration as the main unifying idea. In the other words, we hypothezise that for some developmental successions, at least, it will be enough to know the macromorphology and the mechanical stress pattern of a given developmental stage for deriving from this sole information the similar features of the latter stage, and so on.

This claim requires some immediate comments. In no way does it mean our negation of any non-mechanical (and first of all chemical) factors of development. It means only that we would like to ascribe to the latter factors the well-defined and much respected role of *parameters*, appearing in the morphomechanical equations or in the verbal constructions which replace them. Yes, we know perfectly well, that the parameters, by changing their values, are able to modify immensely a system's behavior, up to a complete elimination of some of its dynamic components. But outside the context of the equations, the parameters themselves are completely meaningless. If we know the equations without knowing the parameters' values we can tell much about a system's behavior (at least in terms of its potencies). On the other hand, if we know the parameters' values without knowing the equations we know nothing.

By following this approach, we may be confronted with the following possibilities. For some cases we may be happy enough to derive the morphomechanics of the next stage from that of the preceding one, without knowing something else. Let us define such developmental periods as *morphomechanically closed*. Another possibility will be that inspite of an obviously mechanical character of a given process, its dynamics will be so much modified by an independent dynamics of chemical parameters, that information about the latter will become indispensable. The systems (developmental successions) of such a kind will be called *mechanochemical*. The next possibility will be that in some crucial moment a closeness of succession, be it morphomechanical or mechanochemical, is disturbed by some unexpected event, not derivable from an embracing law. These systems (or the corresponding developmental moments) will be defined as *opened*. And, lastly, we must be ready to meet some essentially *non-mechanical* systems, the knowledge of their macromorphology and mechanics adding nothing to our ability to reconstruct any next stage of development.

Meanwhile, whatever may be the real situation, our next task will be to define the *relative spatio-temporal dimensions* of the developmental units to be derived from each other. Are they close to the infinitesimal ones, so that an entire developmental course should be described by a kind of differential equation, or, instead, will it be more reasonable to split development into a restricted number of definitely outlined finite macroscopic entities? As many times before, our response will be formulated in terms of levels. While the lower levels' events can be regarded, in comparison with the upper ones', as almost infinitesimal, the latter should definitely be treated as finite. As we hope to show later on, most of the really important developmental successions are definitely associated with the upper levels, that is with the finite spatio-temporal units. In this respect the already mentioned notion of a 'metamoment' (see Introduction, p. 4), joining together a moment of a present time and its finite past into an inseparable entity, will be again useful.

After these comments let us address ourselves to some real developmental successions. First of all, we will give a brief and rather scattered review of the mechanics of the earliest developmental stages (an 'egg mechanics'). Then we will come to some more detailed constructions of the developmental succession, embracing the blastula – neurula developmental period of amphibian embryos (with discourses to other vertebrates). Even later we will cross a large gap and come to a somewhat exotic group of lower invertebrates, namely to hydroid polypes (Cnidaria, Hydrozoa) from Thecate subdivision. Our interest in this group is associated first of all with the fact that it may be considered as a real, and evolutionarily quite ancient 'laboratory of morphogenesis', where some of its main tools, later on to be largely exploited by higher Metazoans, have been first elaborated. Secondly, this group of animals will show us in what way a morphogenesis should be modified for producing a great variety of mutually related species-specific shapes. In the other words, we will explore whether our suggested morphomechanical laws will be robust (structurally stable) enough to be applicable to different species, thus liberating us from the depressing task of building a completely different morphomechanics for any other species.

Our subsequent description may look rather pedantic, including a number of details which are usually omitted in the books of that nature. We hope, meanwhile, that the reader will understand that these minor details are of a real necessity for passing then to some kind of synthetic embracing laws.

3.2. An egg mechanics

In this section we would like to summarize rather scattered data concerning the patterns of mechanical stresses at the earliest stages of the development in order to look for the possibilities to construct for the period considered some embracing morphomechanical laws. The main mechanical events of the early development are as follows.

3.2.1. A PREFERTILIZATION PERIOD

It is already one of the earliest developmental events, namely egg germination, which is associated with an extensive rise of tension within the surface layer (Detlaf, 1977); this, in its turn, is accompanied by a rise in the turgor pressure within an ooplasm (Fig. 3.1 A, B). One of the main factors for the tension increase is a drastic rearrangement of the cortical actin network: the actin microfilaments become gradually realigned parallel to the egg surface, making the latter contractable (Ryabova, 1995).

The cytoskeletal stress field thus established acquires a singularity (break of continuity) on the animal egg pole. This may be visualized, for example, by the microtubular pattern in a sea urchin egg (Isaeva and Presnov, 1990). Not so much the very existence of the singularity, but its solitarity is of interest. It is known from topology that a vector field on a 3-dimensional surface *should* have at least one singularity, but there well may be more than one such point. It is therefore non-trivial that the continuity of the cytoskeletal subcortical field of an egg is disrupted in one point only, this latter coinciding with an animal pole of an egg. Hence there are just the mechanical events which endow an egg with a holistic order from the very beginning of its development.

In those types of eggs which have a non-spherical surface and a substantial turgor pressure inside some more complicated stress patterns, obeying the Laplace law, should be anticipated. Most intriguing examples of such a kind are given by insect eggs, exhibiting a pronounced and regular antero–posterior and dorso–ventral asymmetry (accurately moulded within the ovary). Assuming, according to the Laplace law, that the tension on the surface is inversely proportional to the local curvature (the latter being as a rule decreased from the anterior to the posterior pole and from the dorsal to the ventral side) we should postulate the existence of the postero–anterior and dorso–ventral descending tension gradients (Fig. 3.1 C). Another property of such an egg will be that the circular (dorso–ventral) tensions should be about twice as high as the longitudinal ones are. Could such stress patterns play any role in the oncoming developmental events, associated by recent data with a succession of genes' expressions? In the light of the dominating chemo–diffusional theories, such a suggestion should look as completely unrealistic. The following arguments show, meanwhile, that such a negation may be too rash:

(1) The diffusion of the chemical substances from the nurse cells to the oocyte is itself mediated by the cytoskeletal elements, and in the first turn by the microtubules (Pokrywka and Stephenson, 1991). By recent data, such tension producing and tension bearing factors as the proteins of the actin cytoskeleton also play an active role in the segregation of differentiation factors and mRNA (Nasmyth and Jansen, 1997).

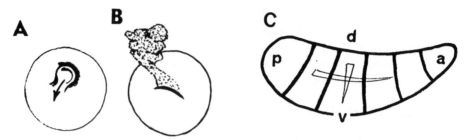

Figure 3.1. Some mechanical properties of eggs. A, B: indications of surface contractility and turgor pressure in a frog's (*Rana temporaria*) egg after its maturation. A: formation and subsequent retraction (arrow) of a protrusion after an egg's incision; this points to an egg's capacity to restore perturbed tensions. B: ooplasmic outburst after incision, indicating turgor pressure within (from Detlaff, 1977). C: anteroposterior (*ap*) and dorsoventral (*dv*) asymmetry of an insect egg. Empty triangles show the expected tension gradients, roughly estimated according to Laplace law. Transversal hoops indicate isolines of *ap* tension gradient. The hoops are arranged similarly to the expression bands of the pair-rule and segment polarity genes.

(2) The cytoplasmic factors which transform, while being locally injected, the eggs of a wild type with a normal antero-posterior polarity into bipolar ones (belonging to the *bicoid* type) abolish, at the same time, the morphological antero-posterior dissymmetry of an egg and a hatched larva (see, e.g., Nüsslein-Volhard *et al.*, 1987). Accordingly, the stress patterns of the transformed eggs also become symmetrical. Is not this the argument for its importance in further embryonic patterning?

(3) The expression stripes of the pair-rule and the segment-polarity genes are arranged not exactly parallel to each other (as one might expect for the case of the purely inhibitory interactions of the neighbouring stripes); instead, they perfectly coincide with slightly diverged and curved lines which correspond to zero gradients of the antero-posterior tensions (Fig. 3.1 C) . For interpreting such an orientation in the terms of a purely chemical signalization a number of additional assumptions is at least required, while the mechanical interpretation is much more straightforward.

3.2.2. FERTILIZATION AND THE FIRST CLEAVAGE CYCLE

The next crucial moment in the dynamics of a stress field onto the egg surface is associated with the eggs' activation (normally occurring in a few minutes after the attachment of a sperm to the egg surface). Now the main event drastically changing an egg's mechanics is an explosive exocytosis of the cortical granules (or, better to say, alveols) leading to an enormous increase in the area of an egg's membrane (Fig. 3.2 A–C). This process as well as its immediate consequences have been studied in much detail on fish (*Misgurnus fossilis*) eggs by Ivanenkov *et al* (1990). In a very schematical way, some of their results are represented in Fig. 3.2 D, E (see also Fig. 2.6 D, p. 93). According to expectations, the authors observed an extensive folding of egg surface taking place in few minutes after fertilization, and obeying a definite animo-vegetal

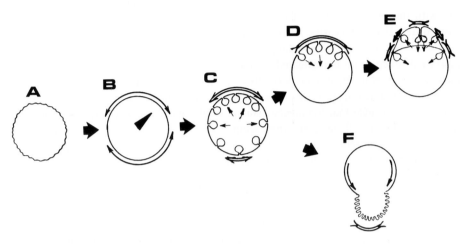

Figure 3.2. A generalized scheme of the stress dynamics on the egg's surface from a prematuration stage up to the completion of the 1st cleavage division cycle. Before maturation an egg surface is somewhat relaxed (A) while the maturation is associated with the tension increase (B). Immediately after fertilization it is hyperrelaxed again because of exocytosis of the cortical granules (C). D, E shows a subsequent regionalization of the endo-exocytotic dynamics and, correspondingly, of the contraction-extension zones during 1st cleavage division cycle in fish eggs (data from Ivanenkov et al., 1990). F indicates the cortical phenomenae (shrinkage of the polar lobe surface) in mollusc eggs (data from Dohmen and May, 1977).

gradient: in 50 min after fertilization the vegetal surface was completely smooth, while the animal one extensively ruffled. An entire egg became at this time pear-shaped, with its maximal curvature near the animal pole, what confirms the existence of a descending vegeto-animal tension gradient.

Such a gradient may be a driving force for the immediately succeeded events, which include: (a) an extensive endocytosis, most of all expressed on the animal egg surface; (b) a concentration of a yolk-free ooplasm just on the animal egg's side. The first event may be considered as a cause, while the second one as a result of the contraction of a previously relaxed animal egg surface. The resulting tension increase seems to overwhelm the initial tension value, since the diminished (because of endocytosis) amount of an egg surface should now embrace the greater area than it did initially. Accordingly, both the endocytosis and the ooplasmic segregation can be regarded as the elements of a hyperrestorative contractile reaction to the surface relaxation caused beforehand by the exocytosis of the cortical alveols. By the similar morphomechanical reasons, another round of exocytosis should be later on expected, directed now towards relaxation of the hyperincreased tensions. This is just what is actually taking place, although if the exocytotic (expanding) zones become now localized only to the lateral from the just initiated first cleavage division furrow. In their turn they become flanked by two endocytotic zones, one of them located in the furrow bottom, while the other is concentrated around the egg's equator, just along the blastodisc boundary (Fig. 3.2 E).

Consequently the early development of a *Misgurnus* egg can be regarded as a succession of 'tensile swings', that is several shifts from relaxation to retension and back again, with the obvious hyperrestorative components. Their T_{ch}-s are of several dozens minutes order. Each successive swing, or, if speaking more pedantically, each next non-linear response to the preceeding morphomechanical situation, makes an egg's organization even more complicated (reduces its symmetry order). The absence of such swings on the vegetal egg surface correlates with its much less morphogenetic activity.

The data obtained up to now on the ascidians' and on the polar lobe-bearing mollusc eggs makes it plausible that the mechanisms of ooplasmic segregation in these species are somewhat similar. In all the cases they seem to be associated, to a great extent, with a postfertilization contraction and a displacement of the membrane and membrane-adjacent ooplasmic components towards the vegetal pole (Fig. 3.2 F). In ascidians' eggs such a contraction, providing a vegetopetal shift of the so called myoplasm has been directly observed (Jeffery, 1992); in mollusc eggs it can be deduced from the extensive folding of the polar lobe membrane surface (Dohmen and May, 1977). Similarly to fish eggs, in both latter cases such a contraction can be regarded both as a hyperrestorative response to the relaxation of the egg surface caused by the exocytosis of cortical alveols and a prerequisite for further developmental events.

3.2.3. MECHANICS OF SAGGITALIZATION

Simultaneously with the above described events, or slightly later, but in any case still during the first cleavage division cycle, most eggs lose their initial rotational symmetry of $\infty \cdot m$ order retaining only a reflectional ($1 \cdot m$) symmetry. In biological terms, this is described as the establishment of a saggital plane (saggitalization) and of the dorso-ventral polarity. More detailed observations indicate as a rule that the reflectional symmetry is not so accurate and an egg acquires at the same time a macroscopic left–right dissymmetry (enantiomorphism), thus diminishing its symmetry order up to 1. In a rough approximation, however, the disymmetry can be often neglected. What is even more important, it is always related to the egg's dorso–ventrality, as to a rigid reference set.

As shown by a series of important recent investigations (Gerhart *et al.*, 1988; Houliston and Elinson, 1991) the saggitalization of an amphibian egg is associated with a microtubules-mediated rotation of a 2-5 mkm thick cortical layer relative to its cytoplasmic core by about 30^0 of arc (350 mkm distance). The plane of rotation coincides with the future saggital axis, and its direction points to the future dorsal side. Actually, this direction is determined by the eccentricity of the position of a spermaster (a centriole brought by a spermatozoon) within an egg: it is a spermaster which plays, within an egg, a role of a microtubules–organizing center, providing thus in a more or a less direct manner the development of an extensive microtubules array in the vegetal egg hemisphere; owing to the eccentricity of a spermaster's position, the microtubules are slightly deviated from the radial directions towards those inclined to a future dorsal side of an egg/embryo. They are probably be inclined also towards left or right, because of the internal chirality of microtubules; by some suggestions (Yost, 1991) this may create a basis of a left-right disymmetry in the Vertebrate embryos. It is supposed that the cortical displacement is caused by the kinesin-like motor proteins, shifting the structural components of the cortical layer towards the external ('+') ends of the inclined microtubules, that is in the direction of an observed rotation (Fig. 3.3 A). A similar microtubules–mediated rotation seems to play a similar role in ascidians eggs, shifting the myoplasm into its final dorsal position (Jeffery, 1992).

In amphibian eggs the main result of rotation is the asymmetrical displacement of a peripheral yolk part: the latter is moved, as a thin layer, animalwards along a future dorsal side, creating in this area (just to the vegetal from α_d angle, Fig. 3.3 B) a so called vitelline wall; at the same time the ventro–vegetal yolk (situated vegetally from α_v angle, same frame) forms a vast bulk. As a consequence, the entire yolk body takes the form of a 'homunculus', its ventro–vegetal swollen region corresponding to the location of a future embryonic head, while the thinned dorso-animal side to that of the tail embryo region. It is just this yolk redistribution that is crucial for the establishment of a dorso-ventral polarity: such a redistribution can be achieved, with the same result, by an egg's tilting in the gravitational field, the saggital plane coinciding now with that of the tilting. The yolk rearrangement is accompanied by the formation of a narrow cytoplasmic swirl nearby the animal pole (Danilchik and Denegre, 1991), which lays parallel to the vitelline wall   (Fig. 3.3 C, arrow). In the absence of the mentioned rearrangements(either as a result of the microtubules' destruction or by preventing an egg's tilting in the gravitational field) an embryo retains its initial rotational symmetry, does not develop any dorsal structures and exhibits later on a slow radially symmetrical gastrulation, leading to a gradual coverage of the vegetal hemisphere by the animal ectoderm. Therefore the rearrangements described are required, most of all, for the formation of the dorsal structures. In what way can they do this?

By Gerhart et al (1988) suggestion it is the physical contact between a vegetal part of the rotating cortex with a more animal level of the subcortical ooplasm, achieved owing to the cortical rotation which triggers some chemical reactions responsible for 'dorsalization'. This view implies the pre-existence of a precise animo-vegetal gradient of some specific morphogenetic substances participating in these reactions. For achieving a desired amount of specificity of the reactions, the degree of a rotation should be also enough precise. However, this is not the case: the rotation dependent dorsalization looks like a fairly robust threshold event. The minimal extent of the rotation providing the dorsalization varies from one egg to another from $0°$ to $30°$, while its variation from $13°$ to $30°$ brings no further changes. That demonstrates, at least, a far going vagueness and/or lack of specificity of the postulated gradients. If the main aim of a given developmental period were to provide a local chemical specification of a relatively small egg region, it might be better reached by some more local chemical mechanisms. What we see instead is, meanwhile, a macroscopic mechanical process, embracing the entire egg. Its consequences should be at least twofold. Firstly, the dorso-animal shift of the cortical cytoplasm at least relaxes (if not slightly compresses) more animal cortical area. According to the hyperrestoration idea this should trigger the active contraction of this latter region (Fig. 3.3 A, double line converged arrow). Secondly (and partly as a consequence of just this contraction) a new mechanical equilibrium will be established within the cortical layer. To reveal its main properties, let us consider the diagramms of the mutually equilibrated tension forces within two triple nodules located on the opposite edges of a yolk phase,

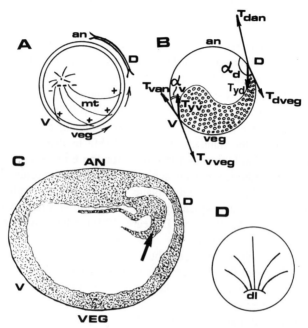

Figure 3.3. Mechanical events associated with the establishment of a saggital symmetry plane in amphibian's eggs. A: microtubules (*mt*)-mediated rotation of a surface cytoplasmic layer in a counter-clockwise direction (from vegetal to the dorsal), single-line arrows. Double-line converging arrow indicates an active contraction of the the animal cytoplasm. B: Force diagramms in the dorsal (D) and ventral (V) nodules. For designations and comments see text. C: convergence of the surface tension lines towards a grey crescent region (a future dorsal blastoporal lip, *dl*)as seen from the dorsal egg's side. D: exact outlines and the arrangement of a peripheral ooplasm (hatched) on the saggital sections of a frog's egg soon after its fertilization. Note the curvatures inequalities between *AN-D* and *V-Veg* regions. Arrow indicates an ooplasmic swirl curled in the dorso-animal direction (data from Danilchik and Denegre, 1991). *an* - animal pole, *veg* - vegetal pole, *d* - dorsal side, *v* - ventral side.

a dorsal (*D*) and a ventral (*V*) ones (Fig. 3.3 B). We assume that in the both cases the main forces involved are: (1) subcortical tensions, and (2) interfacial tensions on the yolk/yolk-free cytoplasm border (as argued by Ivanenkov *et al.*, 1987; Dorfman and Cherdantzev, 1977, these latter are comparable with the subcortical tension forces). As it comes from the nodules' geometry ($\alpha_d > 90°$ while $\alpha_v < 90°$) the oppositely directed subcortical tension forces applied to the same points are of unequal values: $T_{dan} > T_{dveg}$ and $T_{vveg} > T_{van}$. If taking also, as a simplest assumption (equivalent to that of a homogeneity of a yolk phase and a subcortical layer throughout the entire egg) that the homologous forces applied to the opposite nodules are roughly equal to each other (namely, that $T_{yd} = T_{yv}$, $T_{van} = T_{dveg}$ and $T_{vveg} = T_{dan}$) we get

$$T_{dan} > T_{van} \qquad\qquad (20)$$

How can this inequality be maintained if taking into consideration the mechanical autonomy of the egg's subcortical layer? In order to answer this question we should come out of the saggital section and take into mind the possibility of the lateral divergence–convergence of the tension lines. Then the inequality *(20)* can be explained by (and should be regarded as a sufficient evidence for) the existence of a dorso–ventral (*d-v*) divergence of the subcortical tension lines below the animal egg's surface or (what is the same) for the existence of the animally located tension gradient going down in *d-v* direction - Fig. 3.3 D. (The reverse ventro-dorsal gradient to be expected by the same reasons below the vegetal surface is in any case of a much less morphogenetic importance because of the relative passivity of the vegetal area). The existence of *d-v* tension gradient is confirmed by the egg's geometry, namely by the greatest flattening of the dorso-animal region (Fig. 3.3 C, *AN-D* area). Remarkably, such a flattening is maintained up to the early gastrula stage (Danilchik and Denegre, 1991); as we will later on demonstrate, at this very stage it is used as an initial mechanical condition for further morphogenesis. In any case, what becomes emerged soon after fertilization in the vicinity of the dorso-saggital point *D* may be characterized as a powerful singularity of a surface stress field, or, in the other words, as a real mechanical organizer of a subsequent development. Let us recall that it is the second developmental singularity, if taking as the first that established at the animal egg pole at the moment of the egg's germination.

At least in fish eggs the inequality *(20)* may be traced throughout the whole cleavage period and determines probably the dorso-ventral differences in cell's epiboly (Cherdantzeva and Cherdantzev, 1985). A small α_v value on the ventral egg's side predicts the flattening of the corresponding region of the animal cap as well as its smooth vegetalwards spreading. On the other hand, a high α_d value prescribes a sharp segregation of a dorsal ooplasm (and the descending blastomeres) from a more vegetal area, which is just taking place in this very region.

Traditionally the events taking place soon after egg's fertilization are believed to be reduced to a precise redistribution of the cytoplasmic substances ('ooplasmic segregation'). Without neglecting the importance of such a redistribution we would like to note that these very processes may play at the same time another and no less important role by establishing in a regular and a robust manner the holistic stress fields used later on as initial conditions for the subsequent morphogenetic processes.

3.2.4. AN EGG AS AN OPENED CHEMO-MORPHO-MECHANICAL SYSTEM

In spite of large deficiencies in our present day knowledge we may conclude that the mechanical events play a really important role in the dynamics of the earliest development. For several most important events it is even possible to trace, although if in general outlines, a self-derivable morphomechanical succession, probably based upon the abovementioned 'hyperrestoration swings'. On the other hand, such successions are not very long and are initiated from outside, most of all during the fertilization or activation of an egg. As concerning the germination period, a gradual rise of the surface tension seems to be driven most probably, by the shifts of the chemical parameters of the ooplasm. Consequently this initial period of an egg's development can be attributed to what we have defined as the chemomechanical processes. As development proceeds the mechanics seems to be more and more self-maintained, that is, closed within itself.

Similar conclusions can be drawn as concerning the cleavage period. On the one hand, an entire succession of cleavage divisions can well be considered as a closed (self-maintained) series of hyperrestoration swings, its main interrelated components being the endocytosis-mediated contraction within a furrow region and exocytosis-mediated extension on its flanks. On the other hand, one should not forget that each next cytotomy is initiated *de novo* by some factors associated with a mitotic spindle, that is, alien to the abovementioned membrane–bound events. Such an outside regulation of a self-generated succession of events seems to be a rather common case for early development.

Probably only in Spiralians the subsequent cleavage cycles bring something really new in the mechanics of an early embryo: this is exemplified, best of all, by a drastic topological transformation taking place in the eggs of some molluscs at 24 (*Lymnaea*) or 32 (*Patella*) cells stage: at this moment one of the vegetal blastomeres increases the number of its cell neighbours from 6 to 24 (Arnolds *et al.*, 1983). This 'socially active' blastomere will give rise to the main mesodermal derivates. Interestingly, in no way this blastomere had been predetermined to play such a role, so that we are clearly dealing here with the fixation of an occasional choice. With this remarkable exception, the main morphomechanical function of cleavage is restricted by transmitting the stress pattern established at the moment of the egg's saggitalization, up to a blastula–early gastrula stage.

3.3. A morphomechanical reconstruction of a blastula-neurula period of amphibians' development, with some discourses to other taxonomic groups.

3.3.1. OUTLINING THE STRUCTURALLY DYNAMICAL LEVELS

As shown by both localized dissections tests (see pp. 79–80, Fig. 2.1 and the related references) and measurments of the edge angles (see p. 86, equation (15)) this period of amphibians' development can be represented as a succession of mechanical stress fields, each of them remaining topologically invariant during prolonged enough time periods (coinciding with those defined in classical embryology as a blastulation, gastrulation, neurulation, and tail-bud formation) for drastically changing (within no more than a few dozens of minutes) at the beginning of each next period (Fig. 3.4, A–I). What is most typical for each of these fields is the arrangement of so called *cross lines*, that is, the files of stretched cells (shown schematically in Fig. 3.4 by heavy lines) which cross embryonic tissues in the different directions and separate the embryonic body into an increased number of mechanically closed contours. A cellular structure of the cross-lines is illustrated by Figs 2.22 A,B and 2.27 E. The abovementioned drastic topological transformations of the stress fields are characterized, most of all, by the appearance of new 'bundles' of cross lines delimiting the newly established mechanically closed contours. Thus, at the late blastula stage the only detectable cross lines bundles are located within a so-called marginal zone, roughly corresponding to a future mesoderm (Fig. 3.4 A). During a subsequent gastrulation the main changes of a stress field are associated with the increase in its dorso–ventral asymmetry because of the dorsalwards spread of the tensions

(Fig. 3.4 B); no real topological transformations are taking place within this whole period. On the contrary, such a transformation occurs just prior to neurulation: this is the formation of a powerful cross lines bundle in the trunk region coming to the dorsal embryo surface just along the midline (Fig. 3.4 D, F; cf Fig. 2.22 B) or two such symmetric bundles of a similar arrangement in the future head region (Fig. 3.4 E). This may be qualified as a main topological transformations of a stress field during the development of the vertebrate embryos. The established stress pattern is closely associated with the arrangement of the axial organs; meanwhile, the cross lines well may pass the borders of the different rudiments, coming, for example, from the lateral mesoderm to the rudimentary somites and then to the neural ectoderm (see Fig. 2.22B). Thus the cross lines subdivide an embryonic body into a definite number of the stressed compartments which since approximately from the early tail bud stage coincide with the individual organs and body parts such as head and trunk regions (Fig. 3.4 G), gill rudiments, etc.. (Fig. 3.4 H, I).

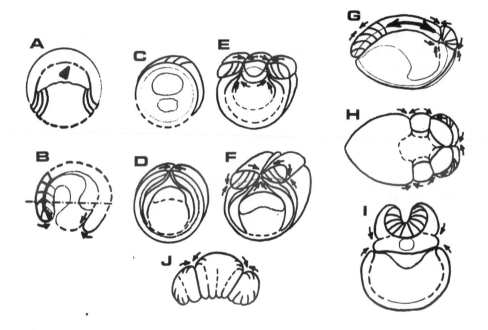

Figure 3.4. Maps of mechanical stresses for several successive developmental periods of a frog embryo (*Rana temporaria*) development. A: late blastula. B: midgastrula. C: same stage, section along the line shown in B. D: transition from gastrula to neurula, posterior region. E: anterior region of early neurula. F: posterior region, same stage. G: early-midneurula,. H, I: mid-late neurula. J: a fragment of ectoderm together with the underlaid mesoderm in 10-15 min after its extirpation from a lateral area of a neurula stage embryo. A, C-F and I: transverse sections, B, G: saggital sections, H: frontal section. Heavy contours indicate the mostly pronounced tension lines, dotted contours - dispersed tensions, fine lines - non-tensed surfaces. Dense double-headed arrow in G shows a longitudinal pressure stress in the dorsal wall (notochord+neural plate). Filled triangle in A is the turgor pressure within blastocoel. Small converged arrows point to the tension gradients in the vicinity of the main cross-lines bundles.

In all these cases the formation of a stress pattern (as revealed by the dissections tests) precedes the visible morphogenesis of the corresponding regions in few dozens minutes at least. The formation of a notochord brings into being a source of a longitudinally directed pressure force within embryo (Fig. 3.4 G, heavy arrow with diverging arrowheads). Prior to this stage the tensile forces have been balanced   only by the turgor pressure in the cavities and in the endodermal cell mass. Fig. 3.4 J illustrates a capacity of any piece of embryonic tissue to create, within few dozens minutes after its isolation a regular stress pattern similar to that observed in the intact embryos.

The formation of cross-lines is a clear example of a contact cell polarization (CCP) in its 'longitudinal version' (see Fig. 2.27 C, E). Meanwhile, at the same time the transversely oriented CCP (Fig. 2.27 D, F) is dominated on the surfaces of the stressed compartments subdividing them, in accordance with the Belintzev's model, into domains of the columnarized (mostly on the dorsal surface of a neurula stage embryo) and flattened (in the latero-ventral regions) cells. In such a way not only the cross lines, but also the embryonic surfaces (and most of all the external layer of the ectoderm, the epiectoderm) become extensively tensed. Meanwhile, what we actually have on the embryonic surfaces are the tensions' gradients rather than uniformly distributed stresses. The greatest tensions values are   exhibited in the immediate vicinity of the cross lines' exits onto embryonic surfaces. The exit areas are shown in Fig. 3.4 B, D – J by small converging arrows.

The above description makes it possible to distinguish two structurally dynamical levels involved to a greatest extent in amphibians' morphogenesis. The lowest one of them may be called a CCP level and corresponds to the level 4 from our general classification (p. 71, 1.5.2). This level's L_{ch} is a minimal distance permitting to register CCP. Accordingly, it is roughly equal to a sum of the diameters of two neighbouring cells ($\cong 10^{-5}$ m). Correspondingly, this level's T_{ch} is estimated as a duration of CCP passage from one adjacent cell to another (this lies in a few minutes range). The next level towards up (corresponding to the level 5 of our classification) is exemplified by embryonic compartments each of them having a topologically invariable stress pattern. Consequently, this level's L_{ch} -s should correspond to the compartments' diameters ($\cong 10^{-4} - 10^{-3}$ m) while T_{ch} -s to their life times (to the periods between two successive topological transformations) that is, to few hours. This amount of time includes: (1) a brief period of establishment of a new stress pattern under the influence of an external force; (2) a memorization of this pattern; (3) a relatively slow hyperrestorative (quasi-relaxational) reaction to an imposed stress. Such a consequence is most of all obvious for the neurulation period, which, as we know already (2.3.6), is characterized by a quasi-relaxation of highly stressed neural plate's cell fan. A same scheme is, however, suitable for other periods as well, including gastrulation, insofar the latter can be also considered as a quasirelaxational contraction of the extensively stretched marginal zone' cells (a plausibility of such an assumption has been proved by modelling (Beloussov and Lakirev, 1988)). Generally, each level's T_{ch} is equal to the duration of an entire HR loop.

A morphogenetic function of each one level can be ascertain by the following simple experiments (Beloussov, 1979). If explanting those parts of an embryo which still do not contain, at a given stage, well developed stress fields the CCP and other related processes will go on in these explants with a considerable intensity but in a rather chaotic way. As a result the explants produce a number of embryonic rudiments each one of them taken separately being normally structured but all of them together arranged in quite an irregular manner (Fig. 3.5 A, B, E). On the other hand, if explanting any regions already acquiring the pronounced stress fields a perfect morphological order will be rapidly established (Fig. 3.5 C, D, F). Hence, the local capacities for CCP and a subsequent cell differentiation are resided onto a lower level while it is the upper one (that corresponding to the macroscopic stress fields) which provides a long-range order. Such a distribution of the morphogenetic functions between two coupled levels is rather typical.

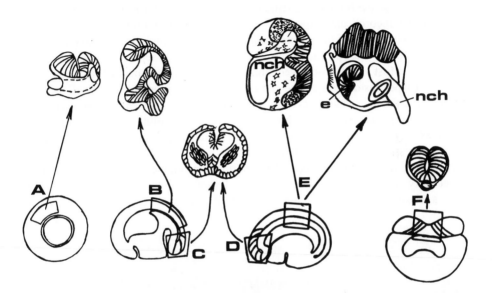

Figure 3.5. Creation and maintenance of a long range morphological order in explanted pieces of embryonic tissues depends upon whether at the moment of explantation the corresponding embryo region possesses (C, D, F) or not (A, B, E) a pronounced tensile field. Tensed structures are shown by heavy lines. e - eye rudiment, nch - notochord. From Beloussov, 1979.

Is it possible now to pass from the above described set of almost static pictures towards a real developmental dynamics? Can the stress fields be considered as some real 'messages' carrying information about a subsequent period of development rather than being mere 'blind results' of what took place before? We have already made the first step on this way by demonstrating that the morphogenesis taking place *within* the limits of each one topologically invariable period can be treated, in a first approximation at least, as a quasi-relaxation of the pre-established stresses. Now we have to make the next step by exploring how these periods can be linked with each other in their natural succession. We hope to do this by using again the idea of the hyperrestoration responses. Let us do this starting from the blastula stage.

As it can be easily tested by the dissections (which we made not only in amphibians, but also in sea urchin embryos: Beloussov and Bogdanovsky, 1980) the blastula wall (the blastocoel roof) of the *Deuterostomia* samples is stretched from inside by the turgor pressure within a blastocoel (in the yolk-rich eggs a similar pressure seems to take place also within a yolky endoderm, which can well be regarded as a mechanically isolated compartment). Relaxation of tensile stresses at blastula stage (by a procedure shown in Fig. 2.26 A, p. 127) led to grave abnormalities. First of all, instead of a progressive thinning of a blastocoel roof (caused by the radial cell intercalation) the number of cell layers within a roof largely increased so that an embryo transformed itself into a solid ball, its blastocoel being shrunk or completely abolished (Fig. 3.6 A, B). Later on an embryonic surface became covered by numerous abnormal protuberances, folds and depressions (Fig. 3.6 C-E). The first order folds or protuberances have been often supplemented by the next order ones so that the fractal–like structures appeared, similar to those shown previously (Fig. 1.3) and theoretically expected (p. 158, 2.4.3). The formation of protuberances indicates a replacement of the normal radial cell intercalation by a tangential one (Fig. 3.6 B, arrows): each one protuberance corresponds to a separate center of such an intercalation. Most of the observed abnormalities can be qualified as those directed towards a (hyper)restoration of the relaxed tensions. Remarkably, most of them can be abolished and the subsequent development essentially normalized if soon after operation the embryos were placed for 4 - 6 hours into hypotonical solution: under these conditions the increased amount of water is pumped into the blastocoel' remnants rapidly restoring the turgor pressure and hence the tangential tensions.

As applied to the interpretation of the normal development, these results demonstrate the necessity of the tensions in the blastocoel roof (maintained by the turgor pressure within a blastocoel) for the *radial* (instead of abnormal tangential) cell intercalation

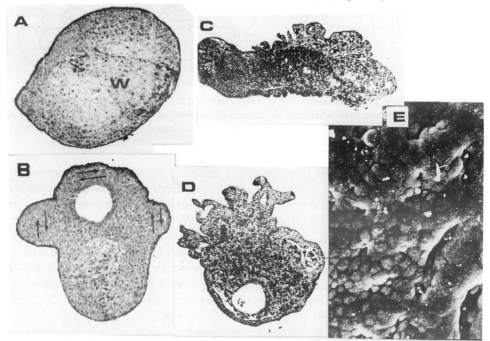

Figure 3.6. Responses of *Xenopus laevis* blastula stage embryos to the tensions' relaxation caused by inserting a wedge of a ventral tissue (w, frame A). A-D: histological sections, E: scanning electron microscopy of a surface area. Converged arrows in B illustrate tangential cell intercalation. D is a cross-section of a protuberance like that shown in C, now indicating the 2nd and 3rd order folds.

within a roof (Fig. 3.7 A, small radial arrows): such oriented intercalation is a natural hyperrestorative response to the increased tensions which leads to the active roof's extension. The immediate consequence of such an extension would be the compression or, at least, relaxation of the neighbouring, more vegetal parts of the embryo, that is, the marginal zone regions (Fig. 3.7 A, B). As a result, because of the action of a hyperrestorative contraction-extension feedback loop (2.5.1) these regions will be actively contracted, promoting further intercalation driven extension of a blastocoel roof, etc. (Fig. 3.7 B). A simultaneous loss of cell-cell contacts in the deep layers of the marginal zone well may be ascribed to the same relaxation-promoted HR reaction (same frame, rounded cells). Let us also recall that the artificial relaxation of a suprablastoporal zone at this very stage was enough for producing several groups of the apically contracted, bottle–shaped cells never seen normally in this location (Fig. 2.36).

The apical contraction, normally adjusted to the initiation point of the blastopore, is the first step of the gastral involution. Its dorsal location is determined as we know, by a dorsal (D) stress singularity established soon after egg's fertilization (Fig. 3.3 B). Meanwhile, even if such an early dissymmetrization were not to take place at all, the circular symmetry of the involution would be in any case broken, because of the competition of the potential contraction zones situated around the blastopore's periphery (Fig. 3.7 C). The domination of the dorsal zone over the flanking blastoporal regions is visualized best of all by the appearance of a dorsal cell fan perfectly seen even from the outside (observation by V. Cherdantzev) – Fig. 3.7 D. In this very case the fan 'predicts' the subsequent encircling and a narrowing of the blastopore. On the other hand, at this very moment or somewhat earlier the blastoporal bottle-shaped cells take the fan configuration in the saggital plan as well, thus 'predicting' the initiation of a gastral involution (Fig. 3.7 B, dotted arch).

This involution will be further reinforced by at least two cooperatively acting feedbacks. The first of them is the curvature increasing feedback (2.5.1., see Fig. 2.41 C) while the other one is E–E positive feedback (2.5.2) established between two adjacent epithelial layers and promoting convergent cell intercalation within each of them. In the vertebrate embryos, with their pre- and post-involuted layers closely adhered to each other, the latter feedback seems to be most powerful of all (Fig. 3.7 E). Its importance for the gastrulation of amphibian embryos has been detected by the separation of both layers without detaching them from an entire embryo: under these conditions the involution stops immediately (unpublished observations). The formation of the axial organs and the neurulation proper may be regarded as immediate consequences of the events described: the gastrulation movements stretch longitudinally the dorsal embryonic surface (Fig. 3.7 F), this initially passive stretching being soon transformed into the active movements of the latero–medial cell convergence directed towards the stress hyperrestoration after just imposed longitudinal tensions. Such an extension inevitably produces two relaxation/compression zones at its opposite poles, the posterior and anterior ones. While the latter one is constantly released by the continuing involution, the anterior zone is, instead, reinforced and transformed thus into a vast region which is relaxed/compressed in the longitudinal direction and, consequently, stretched to some extent transversely (in the frontal direction) - Fig. 3.7 G. This deformation, passive in its beginning, should be later on (during neurulation) reinforced by the active hyperrestorative tendencies, producing

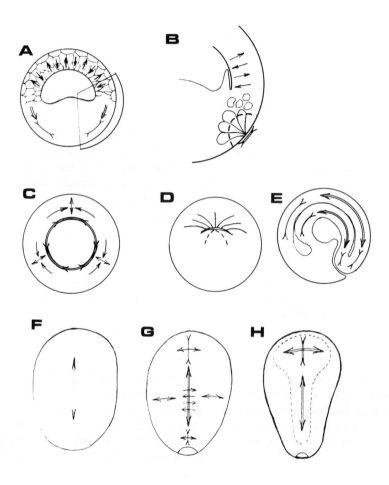

Figure 3.7. Blastula –neurula period of amphibians' development regarded as a succession of the stress hyperrestorations. A: blastula, a saggital view. B: the fragment framed in A at the early gastrula stage. C: a stress pattern created by the active contraction of the blastopore perimeter. The competition of the contraction forces acting in different meridional locations will break, sooner or later, the radial symmetry of the pattern. D: a 'prognostical' fan formed by the cell rows converged towards a just   initiated blastoporal slit. Dashed is the future configuration of the semicircular blastopore, 'predicted' by the fan. E: stress patterns in the advanced gastrula stage embryo, saggital view. F - H: a complication of stress patterns during neurulation, dorsal views. Posterior (blastoporal) pole is to below. In all the frames the double-line arrows show the active stresses while one-line arrows (except those oppositely directed in A, B and G) indicate the passive stresses.

active longitudinal contraction and a frontal (intercalation mediated) extension (Fig. 3.7 H).

Now we can address ourselves to the neurulation proper, that is to the formation and a subsequent rolling of a neural plate. The morphomechanical scheme of this process is, in its general outlines, quite clear and based upon the abovementioned presumptions. Its main points are as follows.

(1) Antero-posterior elongation of a notochordal rudiment will stretch the presumptive neural plate longitudinally and shrink, or at least relax it in the transversal direction.

(2) This shrinkage/relaxation will trigger the active hyperrestorative contractile reaction leading to the columnarization of the neural plate cells. Because of a domination of an apical contraction over the basal one the plate takes a form of the cell fan.

(3) According to Belintzev's model, in the presence of the circular tangential tensions the columnarized cells' domain (cell fan) should be sharply segregated from more ventral areas of flattened cells.

(4) The very rolling of a neural plate into a tube may be considered, as we already know, as a quasi-relaxation of the cell fan.

(5) Formation of a fore-brain is a direct result of the abovementioned active longitudinal contraction and transversal extension of the anterior part of a neural plate.

We have already presented several evidences indicating the patterning and ordering functions of tensions in the morphogenesis of a neural plate (see Fig. 2.26 and the corresponding comments, p. 126). Now we would like to demonstrate, in addition, some more postponed results of the tensions relaxation (Fig. 3.8 A-D). The resulted malformations appeared to be rather various but expectable: a du- and even triplication of the neural tubes (Fig. 3.8 A-C) can be explained by the abnormal ventral extension of a

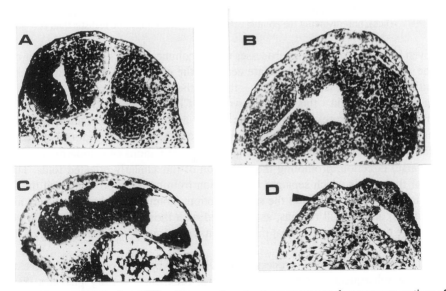

Figure 3.8. Abnormalities in the neural and surrounding structures as seen on the transverse sections of relaxed embryos, 2 days after operation (same as shown in Fig. 2.26). Pointer in D indicates the abnormal fusion of the neural structures with surrounding tissues (from Beloussov et al., 1990).

neural material in relaxed embryos (cf Fig. 2.26 C, D) while a strange fusion of the neural rudiments with the adjacent mesodermal ones (Fig. 3.8 D) by the relaxational inhibition of the delamination tendencies.

A subsequent morphogenesis of a neural tube after its closure can be derived, to a fairly good approximation, from the curvature increasing rule (2.3.7; 2.5.1), if also taking into consideration the anteriorwards wave of the neuroepithelial cells' contraction (as demonstrated by Dr Saveliev; personal communication) and a corresponding wave of a cell–cell pressure increase, going on towards the tip of a notochord (Fig. 3.9 A, B).

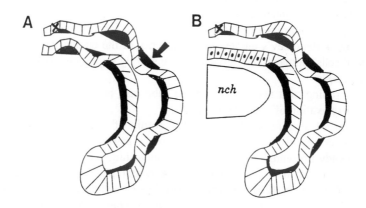

Fig. 3.9 A, B. Modelling of the embryonic brain formation by employing the curvature increase rule. In the both cases the initial contour is black and the resulted (modelled) contour is cross-hatched. Small crosses in the left upper angles show the initiation point of the tangential pressure wave, assumed to move clockwisely around the rudiment's perimeter. In B the wave is suggested to be stopped by the notochord (nch), non-involving the pointed cells. B configuration is more realistic than A.

Even latest morphogenetic processes can be commented on here only in brief outlines. In the vertebrate embryos there are certainly the neurulation movements which mould an entire embryonic shape, including that of the non-neural parts. As related to the amphibians oral field, this was shown by Cherdantzev (1977) who was able to deform the oral structures in a predictable way by distorting the neurulation movements. Without going into much detail, we may point out that the ventro-dorsal movements of the folding neural tube will stretch the latero–ventral regions of ectoderm in the circular (transverse) direction, promoting the formation of the similarly oriented folds. There are some of these folds which separate a gill region from both the posterior and the anterior ones and orient the gill slits in the same direction. Noteworthy, the oral (facial, in higher Vertebrates) and the gill regions, as compared with those more posteriorly and ventrally located are much less tensed. That may explain a subsequent extensive folding of the two first areas.

All of the organ forming processes, such as (may be, first of all!) limb development, but also a subsequent shapening of the neural rudiments, sensory placodes, heart and blood vessels rudiments, etc., form a large set of exciting but still almost non-explored morphomechanical tasks. In any case, regularly arranged mechanical stress patterns are definitely there. In chicken embryos mesoderm they organize cooperative cell movements and shape changes which clearly exhibit CCP properties and participate in somites formation (Beloussov and Naumidi, 1983). Complicated stress patterns in human embryos have been described in the fundamental treatise by Bleschmidt and Hasser (1978). The role of mechanical stresses in the development of a heart loop and the blood vessels have been discovered by Taber *et al.* (1995) and Vaishnav and Vossoughi (1987). In these and other related cases the next aim would be to pass from statics to dynamics, that is, to consider any of the discovered stress patterns as the consequences of the earlier ones and the causes/initial conditions for those later appeared. Such a research trend seems to be very promising not only from the fundamental point of view, but also as regarding its medical and bioengineering applications. It would be reasonable, however, to put it aside until some future date and to discuss instead the following two points: (a) to what extent can the previous analysis of the stress patterns in early amphibian development be applied to the representatives of other related taxonomic groups? (b) what can we tell at the present time about the mechanical components of cell differentiation and embryonic inductions?

3.3.2. SOME DISCOURSES FROM AMPHIBIAN EMBRYOS TO THOSE OF OTHER RELATED TAXONOMIC GROUPS

A suggested morphomechanical scheme of amphibian development would not certainly be of a very much interest if some of its main principles could not be applied to the representatives of the other taxonomic groups. Happily, this is not the case: some main points seem to be applicable even to most of all remoted species. In a broadest aspect, the processes similar to contact cell polarization seem to take place in all the Metazoa, with a possible exception of Sponges. In any case, they are clearly visible during bud formation in hydroid polypes (p. 124) as well as during their embryonic development (Krauss and Cherdantzev, 1995 and personal communication). A mostly fundamental distinction of this group, together with other lower invertebrates, from the higher animals is meanwhile the lack of any higher levels' stressed structures during *embryonic* life period (although there is an entire peculiar hierarchy of such structures in their vegetal generations (see 3.4).

Some important evolutionary acquisitions are associated indeed with the extension of the regular stress patterns towards a macrolevel. One of the main milestones on this way was the 'invention' (most probably, by some ancient Deuterostomia) of a *blastopore with the actively contracting periphery* (as shown on Fig. 3.7 C). On one hand, the active peripheral contractility is an easy way for maintaining a circular shape of a blastopore during its progressive narrowing. And, on the other hand, such a maintaining would be mechanically impossible (or, better to say, very unstable) if trying to shrink a *passive* blastopore from outside with the use of a certain planar pressure, for example by the epiboly associated with cell proliferation. Any attempt to do this (easily reproducible on sheet of a paper) will immediately produce a lot of the radial folds at the blastopore periphery. Then, a tendency for the energy minimization will drive a   deformed hole towards the elongated shape, producing thus most probably a slit like blastopore instead of a circular one. This is, we suggest, the main reason for the ubiquitity of a slit like blastopore in Protostomia embryos, assuming that the latter widely use the epibolic mode of gastrulation and did not yet 'invent' the mechanisms of a circular contractility.

At the same time, the formation of a blastopore with an actively contracting periphery will very much amplify the contraction–extension feedback loop (2.5.1), which can now extend its action over the entire embryo: the embryonic surface will respond to blastoporal contraction by its active extension, mediated in the multilayer sheets by the radial cell intercalation and in the monolayered sheets by the cells' flattening. Certainly, this feedback will be active only if the entire construction is under tension. As we know, this is almost always the case, first of all, because of a turgor pressure within a blastocoel. Although, at first glance, the turgor pressure force is opposite to the involution tendencies, the latter are on the contrary very much inhibited and/or distorted in the case of a pressure *decrease* not only in amphibians, but also in sea urchin embryos (Beloussov and Zhadan, 1993).

The Echinodermata embryos already possess all that is required for the main feedback loop (2.5.1) to come into action. The main distinction from the Vertebrate embryos is their restricted capacity for epibolical extension (because of impossibility of a cell's repacking within a monolayered blastula wall) and the lack of mechanical contacts between the pre- and post-involuted cell material: as a result, the extension-extension feedback (2.5.2), so important for Vertebrates cannot play here any role. As a consequence, the dorso–ventral dissymmetry remains in Echinodermata gastrulae quite rudimentary and nothing similar to the axial organs of the Vertebrate embryos can be formed.

Meanwhile, the already involuted cell material (the primary gut wall) of Echinodermata embryos is capable of cell intercalation (Hardin and Cheng, 1986; Keller and Hardin, 1987). Coming from our morphomechanical presumptions this should be possible only if the gut rudiment have been initially stretched. Probably, such a stretching is mediated by the fillopodial protrusions which are known to attach the gut tip to the inner wall of a blastocoel (Gustafson and Wolpert, 1967). On the other hand, the gut rudiments are no less extended in the exogastrulating embryos although such a mode of stretching becomes now impossible. Could a stretching force be produced in this case by a turgor pressure within a common blastocoel + gut cavity? This question remains still unanswered.

Before approaching the Vertebrates proper, let us make some remarks about their possible evolutionary ancestors (or, may be, a side branch?), the Hemichordata. An amazing similarity can be traced between these, often proboscis-bearing and very much 'fractalized' (see Fig. 1.3 B) creatures and the monsters arisen from the relaxed amphibians blastulae (Fig. 3.6 C, D). One may speculate that these fractalized shapes are, so to say, very much generic, at least for lower Chordata. In this case, a passage towards real Chordata should be associated with these tendencies surmounting, because of the appearance of the powerful holistic stress fields, which are fundamentally the same in all the Vertebrates embryos. Let us comment this by discussing briefly the morphomechanics of the fish (Teleostei) embryos and of a chicken embryo.

In Teleostei embryos, in spite of their involution being quite rudimentary, if at all present (Solnica-Krezel *et al.*, 1995), the gastrulation driving feedback seems to be very much the same as in amphibians. The main tension-bearing contour includes here the external layer of the ectoderm (an enveloping layer, EVL) being in a direct tangential contact with the external yolk syncytial layer (E-YSL) directly connected with yolk syncytium proper (YL) - Fig. 3.10, top frame. From the blastula stage on, this tension is,

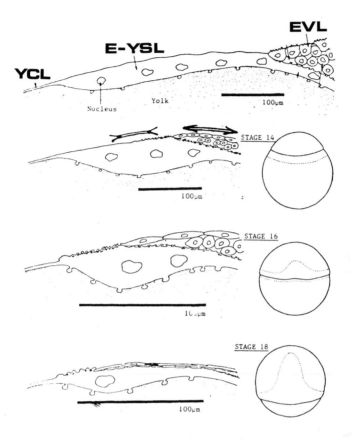

Figure 3.10. The edge of a *Fundulus* egg blastoderm (Teleostei) at four successive stages of development (from top to bottom). The external yolk syncytial layer (E-YSL) together with the underlaid yolk is actively contracted due to endocytosis while the enveloping layer (EVL) is actively expanded due to the radial and the convergent cell intercalation (double-line arrows above the second frame). Thus, the E-YSL behaves similarly to the marginal zone, while EVL similarly to the suprablastoporal region and a blastocoel roof of an amphibian embryo (from Betchaku and Trinkaus, 1978, with the additions and the publisher's permission).

as usual, supported by the turgor pressure within a blastocoel and a yolk sac; similarly to amphibians, the EVL stretching may promote firstly the radial and then the convergent (directed towards a dorsal midline) cell intercalation. Although the position of the dorsal midline is normally predetermined during ooplasmic segregation, it may be reoriented during cleavage period by remodelling a stress pattern (Cherdantseva and Cherdantsev, 1985).

Probably the most interesting point of an entire gastrulation mechanics of a fish embryo is that the EVL and E-YSL are coupled together within the same contraction–extension feedback loop (5.3.1) as are the blastocoel roof and the marginal zone of the amphibian embryo: while EVL is actively extended (by means of either radial or a convergent cell intercalation), the E-YSL is contracted as a result of the ruffling and of the above mentioned (p. 94) endocytosis of a plasma membrane(Fig. 3.10, three lower frames). This feedback loop seems to be the main driving force of the epiboly, permitting the latter to go ahead even under increased resistance while the blastodermal edge crossing an egg's equator: for a feedback of this kind, the more tension the better!

Lastly, we come to the Amniotae, taking the chicken embryo as an example. An indispensability of the tensions within its blastoderm for maintaining the normal development is quite obvious (Bellairs *et al.,* 1967). Interestingly, the first event to be observed in the videofilms of an explanted blastoderm at the beginning of its incubation, is its self-retension caused by the contraction of the *zona pellucida* (Kucera and Monte-Tschudi, 1987). Thus, the establishment of tensions looks as a precondition of a further development. Remarkably, in chicken embryo, as opposed to amphibians, the tensile stresses are maintained only under the normal incubation temperature and are rapidly self-relaxed in the cooled embryos (Beloussov and Naumidi, 1977). Hence, the elastic component of the tension appears to be here inseparable from the active one

A morphomechanical scheme of an early chicken development has fundamental similarities with that of amphibians, and may be represented as follows (Fig. 3.11). The location of a posterior embryo pole is determined somehow by the yolk's redistribution in the gravity field (Kochav and Eyal-Giladi, 1971), this event probably being similar to the yolk shifts within a

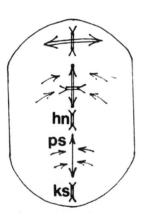

Figure 3.11. Consecutive stress patterns in the blastoderm of Amniota embryos, brought together within a single sketch (temporal succession going on in an upwards direction thus coinciding with the postero-anterior direction). *ks:* Kollar'sickle, *ps:* primitive streak, *hn:* Hensen's node, *tr:* trunk region, *h:* head region. The stresses in *ps* region are indicated as passive ones although later on they become active.

freshly fertilized amphibian egg. A cell immigration at the posterior pole (a formation of so-called Kollar's sickle) indicates some amount of surface contraction within this area. This contraction may trigger, as a result of a feedback loop described in 2.5.3, convergent extension movements taking place more anteriorly. That leads to the formation of a primitive streak. As a result of its active elongation, a new compression/relaxation zone is formed at its anterior end: this is Hensen's node. A subsequent active (hyperrestorative) contraction of the latter should initiate another similar feedback loop located more anteriorly: what it produces now is a longitudinally extended neural tube (hindbrain) and a relaxed/compressed part of a blastodisc ahead of it. The latter gives the forebrain and also the anterior (transverse) amniotic fold (which is homologous to the oral field structures of amphibian embryos). What we see in Amniotes is, hence, a twicely repeated action of the same, most commonly used contraction – extension feedback loop. What may be regarded as an evolutionary novelty of the Amniotes as compared to Anamnia, is just the first round of the feedback interactions (associated with the primitive streak formation), while the second one, associated with the formation of the axial organs, is completely homologous to that in Anamnia. From such a viewpoint, the evolutionary acquisitions in Amniotes should be better evaluated as an insertion, rather than as a classical Haeckel's anaboly.

3.3.3. DOES MORPHOMECHANICS PARTICIPATE IN CELL DIFFERENTIATION AND EMBRYONIC INDUCTIONS ?

Are there any evidences that not only the morphogenesis in its strict sense, (namely, the movements and deformations of the more or less identical cells), but also the intrinsic cell differentiation (closely associated with the expression of specific genes) is affected and even sometimes determined by mechanical factors? Numerous data obtained within the last decade or so, makes such a suggestion pretty well substantiated, at least for some terminal cell differentiations. It turned out indeed that the mechanical stresses, either constant or periodic directly affect cell differentiation at the transcriptional and posttranscriptional levels in a wide range of different cell types: fibroblasts, chondroblasts, osteoblasts, myocytes, endothelial cells, retinal cells, and others (reviews: Opas, 1994; Ingber *et al.*, 1994; Jones *et al.*, 1995; see also Larsson *et al.*, 1991; Rosales *et al.*, 1990; Toshiaki and Bauer, 1992; Korver *et al.*, 1990; Mochitate *et al.*, 1991). In the *Arabidopsis* plant even a rain-, wind and touch-induced pressure appeared to be enough for inducing the expression of calmodulin and calmodulin-related genes (Braam and Davis, 1990)! These data insistantly invite us to explore the intercellular pathways able to transmit mechanical signals from a cell membrane directly to the nuclear matrix, most probably via the cytoskeletal structures (Forgacs, 1995; Traub and Shoeman, 1994). Here we enter the exciting and up to now just slightly touched area of interplexed micromechanical, electrical and energy transfer events taking place within the molecular and supramolecular structures of a living cells spending obviously the main part of their life times far removed from a thermodynamical equilibrium (Aon *et al.*, 1996; Bistolfi, 1991; Fröhlich and Kramer eds, 1983). All of them together create what Goodwin (1985) defined as a common 'chemomechanoelectrical field', and it would be quite strange if the transcriptionally active parts of a cell's genome were not be the integrated parts of such a field.

Returning meanwhile to the problem of embryonic differentiation in *sensu stricto*, we can see that up to now almost in no cases the morphomechanical approaches, in any of their aspects have been used here to a noticeable extent. Instead, patterns of embryonic cell differentiation are widely believed to be based exclusively upon the diffusion of the chemical 'morphogenes and such a viewpoint is fully dominating in the present day studies of embryonic inductions, occupying a central position in at least Vertebrate embryology. These studies belong to rapidly developed and mostly flourishing trends of the present day biology, so that any criticism made from general positions may sound here as a grumbling of an outsider whose arguments, even if looking substantiated, can be easily surmounted by a next round of experiments. I hope to show meanwhile that this is far from being the case. One of my main arguments is that within more than 70 years history of this research trend all the efforts to derive the final morphological results of the inductions from the 'nature' of the inductors themselves has been confronted with the same kind of difficulties which are inspite of all the experimental progress up to now non-surmounted.

The permanency of these difficulties is illustrated, best of all, by a persistent reiteration of the main explanatory constructions suggested in this field since Spemann's times. Actually, we can distinguish here two viewpoints, the first of them to be denoted as a 'strong' version of the inductors' action while the second as a 'weak' one. According to the strong version (as formulated in modern terms) the inductive centers are the sources of some kind of a positional information (PI) which is spread throughout embryonic space as a diffusional field of the inductive substances. It is the cell's position in this field which unambiguously determines its fate in a concentration-dependent manner. One of the first concept of such a kind was a well-known 'double-gradient model' (Saxen and Toivonen, 1962). It implied the existence, at an early gastrula stage, of two inductive centers, a neuralizing and a mesodermalizing ones which were assumed to create together an overlapped diffusional field determining the overall pattern of the axial organs. A next version of the double gradient model (e.g., Slack, 1994) considered both the Nieuwkoop's organizing center (situated at the vegetal pole of a blastula stage embryo) and the classical Spemann's organizer (dorsal blastoporal lip) as two independent PI sources. More recent constructions taking into consideration the epidermalizing factors from so-called BMP group (see, e.g.Graff, 1997), are of the same explanatory structure. The wide spread views upon the double-centered structure of the Spemann's organizer itself (consisting of the trunk– and head–organising activities) may be also attributed to the same category. All the strong versions are common in non-allowing (although in most cases tacitly) any noticeable role in the differentiation and patterning of the axial organs to the processes other than the diffusion of the inductive substances and the concentration-dependent genes expression. The morphogenetic movements, together with the associated stresses are regarded, within this conceptual framework, as no more than the blind "end-results" of the inductive and genes expression events, being themselves unable to affect the latter ones in a reverse manner.

Among the arguments supporting such a viewpoint *in vitro* experiments indicating the concentration-dependent activation of certain genes are often mentioned (Gurdon *et al.*, 1994). However even in these largely simplified models of embryonic differentiation the resulted spatial patterns are much less precise than those achieved during normal development. Much more general objections can be also raised against the strong versions. The matter is that the mechanisms allowed by any of the latter are essentially non-robust: any mutual displacement of the inductive centers (which correspond to the previously discussed 'privileged points', see pp. 36-37) should largely distort the resulted patterns of the axial organs. Interestingly, none of the strong versions makes even a slightest attempt to predict the character of these displacement-caused anomalies, what should be the immediate task of these theories, if taken seriously. And the actual results of the displacements experiments contradict directly to any expectations based upon the strong versions: even far-going displacements, including a complete rearrangement of the inductors' material (see 2.3.5, p. 126) and/or a removal of one of the inductive centers either do not disturb the resulted patterns at all or disturbes it in the tensions-dependent way so that the abnormalities obtained can be abolished by a proper restoration of the relaxed stresses (see p.180).

As to the 'weak versions', their most cautious formulation is that the inductive substances initiate and affect somehow the development of the axial organs without specifying their patterns and the morphological structure in all the details. Thus, according to the 'weak versions' the inductive substances do not create themselves any detailed prepatterns. The pattern formation role is redirected, by this view, to some later processes among which cell rearrangements and the associated events such as changes in cell contacts and the mechanical stresses should play an important and an autonomous role.

A good number of classical concepts of embryonic inductions can be related to weak versions. Such are undoubtedly Waddington's and Needham's ideas about two steps of an inductive process, evocation and individuation, only the latter one establishing precise patterns (Waddington, 1945). Nieuwkoop's activation - transformation model also implies two similar steps, the first of them, namely the activation looking to be rather 'smoothed' and spatially imprecise. Spemann himself fluctuated between the both versions: liking to regard inductors as field sources (and creating thus a germ of a strong version) he had at the same time enough insight for evaluating properly his own experiments on the oral field induction as the indications of an 'abstract' (that is, lacking any definite prepattern) phase of inductive events (Spemann, 1936, S. 235-237).

In more recent times this investigatory trend has been prolonged by Yamada (1981, 1994) who claimed overtly that the problem of induction induction should be treated in terms of the 'morphogenetic events' rather than 'substances'. He argued, in particular, that the mode of action of a primary inductor should be mediated by the morphogenetic movements within either inductor itself or a reacting tissue. For example, if a prechordal plate (an anterior part of a primary inductor, which normally induces forebrain structures) undergoes, contrary to its normal potencies, the movements of a convergent cell intercalations, it induces the hindbrain parts instead of the forebrain. Similarly, a forebrain rudiment, if being affected to produce autonomously the same kinds of movements, will contribute, contrary to its normal fate, to the hindbrain parts. In accordance with this are Sokol's (1994) data showing that the arrest of the intercalation movements in induced explants prevents their expected trend of differentiation as well as the earlier report by Symes and Smith (1987) that one of the first indications of the effect of an inductive substance, the activin, is the initiation of the convergent intercalation cell movements.

Several sets of experiments supporting a morphogenesis-dependence of the axial organs development have been performed in our research group. We could show, for example, that a set of the axial organs developed from an embryonic tissue piece (this may be a neuroectoderm taken from midgastrula stage, or the lateral and even the ventral part of a marginal zone) is increased if this piece is allowed to involute by one of its poles, irrespective of the latter's origin: the involuted part contributes as a rule to the notochord and the axial mesoderm (Beloussov and Petrov, 1983; Beloussov and Snetkova, 1994). On the other hand, the arrest of the involution movements within the dorsal blastoporal lip largely reduced its capacities to produce axial organs (Beloussov and Snetkova, 1994). Also, a very brief (30 s) disaggregation of a dorsal lip tissue in Ca^{2+}-free solution, without any additional inductive influences shifted a set of the produced structures from those specific for the trunk region (somites, notochord) to the fore-brain ones (Georgiev and Beloussov, 1986).

In another series of experiments (Yermakov and Beloussov, 1998) we blocked at the early gastrula stage the dorsalwards convergent intercalation movements in the lateral parts of the marginal zone by making two symmetrical longitudinal incisions which separated these parts from the dorsal blastoporal lip. As shown by morphometrical analysis, the *average* amount of all the kinds of axial tissues (neural, notochordal and somitic) was the same in the operated and the normal embryos; hence, in the average the inductive processes have been neither stimulated nor inhibited by the operations performed. Obtained meanwhile by these interventions was an extensive loss of precision in the spatial patterns of the axial organs: they became highly asymmetrical and variable. One could observe, for example, a vast notochord in one lateral lip and nothing of this rudiment in the opposite one; same was true for somites and the neural derivates. In addition, the volumes' relations of the notochordal, somitic and neural tissue within the different experimental samples were largely abnormal and each time different. These results learn us that the induction *per se* and the establishment of a precise pattern of the axial organs (including their mutual proportions) are essentially different processes, the first of them in no way providing automatically the second one. What is really established by the induction *per se* can be best of all characterized as *a largely delocalized field of the intermixed cell potencies*: none of the cells situated in this field is definitely specified, and no prepatterns at all are as yet established (this corresponds exactly to the Waddington's evocation phase). Certainly, our data are not the only one, supporting such a view: "at the level of individual cells, the initial specification of tissue type in the marginal zone is locally stochastic, incomplete, and reversible. Further local communications seem to be required for the coherent, region-specific differentiation of the various sectors of the marginal zone" (Wilson and Melton, 1994). Similar conclusions may be also drawn from Vodicka and Gerhart (1995) studies as well from the data concerning so called 'community effects' (Gurdon, 1988). The transformation of such a field of potencies into the definite precise patterns endowed by both the short range and long range order (what just corresponds to the individuation phase, by Waddington's terminology) necessarily demands, by our view, morphogenetic movements and the associated mechanical stresses.

From the view of a self-organization theory the inductive events can be represented, in their main outlines, by the model shown in Fig. 1.16 (p. 45). As it we have already mentioned, this model desribes, firstly, a parametrically-driven arising of a competence (that is, the increase of a set of potentially achievable stable states) and, secondly, a dynamically-driven (triggered) selection of a final state out of this set. Most probably, the inductive factors play either role, that is, some of them act as more or less long-termed parameters (providing the arising of a competence) while others as brief triggers promoting the achievement of the final state out of a set provided. Such may be, for example, a role of BMP-like factors in selecting between a skin or a neural developmental pathway (see, e.g., Hemmati-Brivanlou and Melton, 1997). In no cases, meanwhile, the finally achieved structures can be exhaustively derived from either a parameter's or a trigger's 'internal nature' as seen outside the context of an entire system. That this context includes morphomechanics seems to be quite obvious. If only a small part of the up to now essentially reductionistic efforts to reveal the inductive mechanisms were redirected to more global morphomechanical aspects our knowledge of this fundamental and very much interplexed set of events were much more complete and balanced.

3.4. Developmental successions in Thecate hydroids (Cnidaria, Hydrozoa)

3.4.1. A SURVEY OF LEVELS

One of the main advantages of Thecate morphogenesis is that its dynamic levels, although if hardly homologous to any of those above outlined, are quite distinct and may be clearly separated from each other. Among these the most fundamental and well-known is the level of the growth pulsations (GP), its L_{ch} (as measured by the growth advancement per single GP) being of several (less than 10) micrometers scale and T_{ch} (a duration of a single GP period) of several minutes scale (for $O.$ $loveni$ from the White Sea it is, under the optimal temperature, from 10 to 5 min: Beloussov et $al.$, 1989; Kossevitch, 1990). As to non-differentiated horizontal outgrowthes (stolons) which are simply spread over the substrate this is the sole level to be definitely distinguished. Meanwhile, in the vertical outgrowths (peduncules), terminated by the feeding individuums, the hydranths, at least two additional upper levels can be traced. As viewed in an ascending order, those are:

– a level (or a set of closely situated levels) involved in the peduncule's 'sclupturing'. The main 'sculpture elements' are so called annulations, that is the bulbous segments, each of them being built by 10 – 12 successive GPs. Consequently, for each successive annulation $T_{ch} \cong 50$ –120 min and $L_{ch} \cong 50 - 70$ mkm. The annnulations almost never are solitary; instead, they are collected into series, each one of them consisting, in $Obelia$ peduncules, most frequently of 4 units. Correspondingly the T_{ch} and L_{ch} for the annulation series should be multiplied by 4. Lastly, the annulation series are separated by the smooth (non-annulated) peduncule's parts having approximately the same T_{ch} and L_{ch} as the series themselves (Fig. 3.12).

– a 'branching level', embracing the whole cycle between the very initiation of a new peduncule and its termination by the differentiated hydranth. By Kossevitch's (1990) data this cycle corresponds in $O.$ $loveni$ to 140 – 180 GPs (it is of interest that this value is quite stable and does not depend upon the feeding conditions, or the amount of the cell material involved in the peduncule's formation). Correspondingly, for the 'branching level' $T_{ch} \cong 700$–1800 min and $L_{ch} \cong 700$–900 mkm.

In our subsequent account we shall pay main attention to the temporal, rather than spatial components of Thecate morphogenesis. This is closely associated with our above formulated view that the primary targets of any developmental parameters are the rates of the processes, that is the temporal components. As to the spatial structures, we shall consider them as the derivates of the temporal ones, rather than primary autonomous entities. The Thecate samples are most of all suitable for such a kind of studies, as providing a perfect correspondence between the temporal and the spatial patterns: it is a real $dynamic$ architecture what we see here!

Now let us study the above enumerated levels in more detail, starting from the uppermost one.

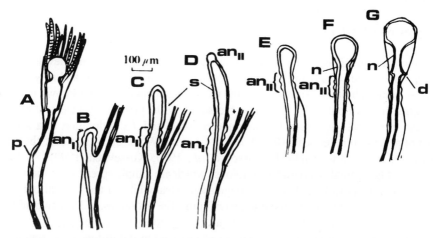

Figure 3.12. The development of an *Obelia loveni* peduncule. A: the initial plateau (*p*) stage on the left side of a maternal stem, culminated by an adult hydranth. B: formation of the 1st annulation series (*an$_i$*). C, D: initiation and elongation of a smooth (non-annulated) zone (*s*). D: beginning of the 2nd annulation zone formation (*an$_{ii}$*). E-G: formation of a hydranth . *n*: a hydranth's neck, *d*: diaphragm. From Beloussov 1973.

3.4.2. A BRANCHING LEVEL

As has already been briefly mentioned, in all of the Thecate hydroids each new peduncule commences its development from the formation of a columnarized cell domain out of an initially flat ectodermal epithelium (Fig. 3.12 A, *p*). The columnarized cells soon elaborate an extensive vacuolar system and start to pulsate. Similar processes start to go on in the underlying endoderm. In general, a peduncule's rudiment just initiated consists entirely of a pulsatorial zone, which starts to produce a distalwards progressing tube, that is a peduncule proper, by the mechanisms described in 2.3.7 . During this movement the more proximal regions of the flattened non-pulsating epithelium are gradually involved into the peduncule, so that proximally to the apical pulsating zone (which always remains to be closely adhered to the just produced perisarc) another one becomes gradually formed. Its cells do not pulsate and are deadhered from the perisarc, looking like a flat epithelium. We shall define the apicalmost pulsatorial zone as the A-zone, while the proximal non-pulsating one will be called the B-zone. As mentioned before, in the peduncules a substantial part of A-zone becomes annulated. This seems to be crucial for the subsequent differentiation of the terminal hydranth: the stolons which lack the annulations do not produce hydranths, and if just one annulation have been produced by a young outgrowth of an unknown origin you may be sure that later on it will become terminated by a hydranth. The latter is always formed from the distalmost annulation of the second annulation series (Fig. 3.12 D-G).

How can this sequence of events be reproduced for an indefinite number of times in such a perfect order? The classical approaches to the pattern problem, associated with linearistic concepts of gradients, dominance, and positional information pay little, if any, attention to the temporal components of the processes under study. One of the widespread beliefs of this kind is that a morphology of each successive proximo-distal level of a daughter stem can be regarded as a function of its distance from the maternal stem (a function of its "positional value").

Some elegant recent experiments (Kossevitch, 1996) showed, meanwhile, that in *O.loveni* at least the developmental fate of each site of a stem is far from being unambiguously determined by its distance from the maternal stem. Instead, quite another mechanism is in action. It implies some kind of a 'distributed memory', located mostly in the tip of the growing peduncule. So, if we remove an intermediate part of a peduncule which had already created the second series of annulation and transplant its tip onto the stump, the transplanted tip will produce just this part of a peduncule as scheduled normally, so that a hydranth will be formed on a shortened peduncule (Fig. 3.13 A). Similarly, if we transplant a 'young' tip to a distal part of a more advanced peduncule, the tip will pass through all of its predestined developmental route, so that a resulted peduncule will be much longer than normally (Fig. 3.13 B). Remarkably, the reversion of an intermediate stem part gives no effects at all (Fig. 3.13 C). On the other hand, whatever be the age of a peduncule, after the removal of a tip the stump will start its development from the very beginning, as if forgetting completely the route already passed through (Fig. 3.13 D-F). To this we may add that an apical region of a newly initiated peduncule isolated from a maternal stem perfectly passes all of its developmental route even if the amount of its cell material is not enough for filling the entire perisarcal tube (Fig. 3.14 B1, D, F, G). Taken together, these experiments show that neither the distance of a given peduncule's level from a maternal stem, nor the peduncule's polarity or the amount of its cell material, but, instead, some kind of a 'memory' about a passed developmental route, located in the tip region, is crucial for deciding what kind of structures will be formed. On the other hand, the stolons do not have such memory at all.

Meanwhile, an isolated peduncule's tip, if it do not possess a certain amount of a cell material in its more proximal parts, can 'forget' its developmental destination and produce some abnormal structures (Fig. 3.14 B 2 and 3, E), while the addition of such material will improve its memory (Fig. 3.14, cf. B 2 and 3 with A, E, F). That means, that a developmental memory, although being mainly located in the tip region, requires also, for acting properly, no less than some critical amount of cell material: it does really look like a delocalized (distributed) one. What morphogenetic events may be associated with such a kind of memory?

It may be most of all important, that during the peduncule's growth (contrary to that of the non-differentiated stolons) the length of the A-zone gradually diminishes, while that of the B-zone is increased at its expense. This means that the most proximal cells of the A-zone are gradually migrating from there to the B-zone. When the A-zone is almost completely exhausted, the growth of a given peduncule stops and the hydranth becomes fully differentiated. We can tentatively conclude, that it is a gradual exhaustion and shortening of A-zone which is 'remembered' by the peduncule's tip and plays a role of a 'branching clock', determining an overall duration of a peduncule's growth and hence its

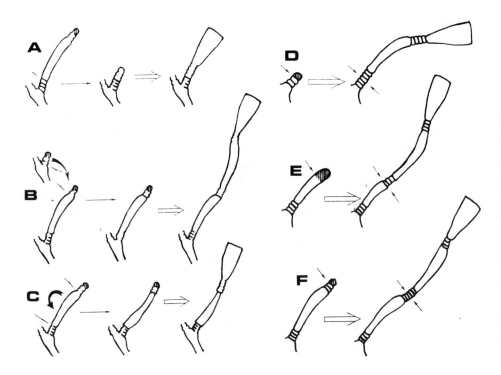

Figure. 3.13. Experiments indicating the 'developmental memory' of a peduncule's tip and the absence of such memory in the stumps of *O. loveni* samples. A: 2nd annulation series is transplanted onto the tip of the 1st annulation series. B: the reverse situation. C: a smooth zone of a stem is inverted. D - F: results of the excisions of the dashed zones shown in left frames. In A-C the developmental fates of the transplanted tips are the function of their own age and do not depend at all from that of a stump. Meanwhile, in D-F experiments a stump always starts its development from the very beginning. From Kossevitch, 1996, with the author's permission.

final length. Now let us come to more details of the processes described.

Let us firstly ask ourselves, what are the factors of the B-zone elongation. To a certain extent this may be a passive process: the entire B-zone is stretched by the actively growing (by means of GPs) A-zone and, also, the B-zone ectoderm may be stretched by the same zone endoderm, and vice versa. However, the passive component is not the only one to be into play. There are several evidences indicating that this very passive stretching triggers, according to the familiar morphogenetic feedback (2.5.1) the *active* intercalation–mediated extension of the stretched B-zone. The main arguments for the existence of such a *stretch-promoted cell intercalation* (SPCI) in the hydroids stems are as follows:

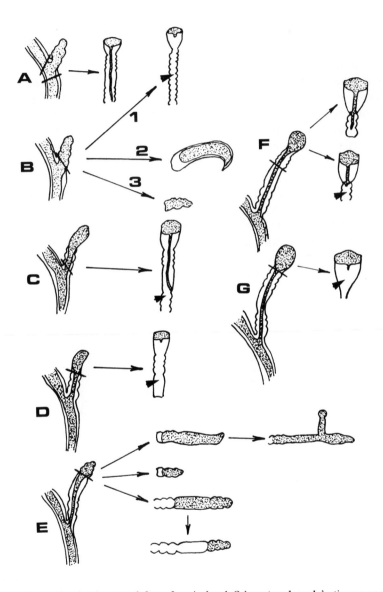

Figure 3.14. The developmental fate of an isolated *O.loveni* peduncule's tip segment largely depends upon the amount of cell material included into it. If an excised piece is itself enough large and/or contains the material of non-annulated regions (A, C, D, F, G) the tip exactly passes its destined developmental pathway and in any case forms a properly shaped hydrotheca (even if its most proximal levels are not filled with the coenosarc - see pointers). In other cases a tip development is rather variable and abnormal (B2, B3 and E).

1. The B-zone elongation is associated with its narrowing, that is with a decrease in the number of the vertical cell rows and an increase in the disto-proximal length of each such row. In Thecates such a narrowing is easily observable because the initial width of a given coenosarc region is firmly 'frozen' by the perisarc. Therefore the narrowing and the corresponding redistribution of the cells can be estimated by the volume of empty space that appears under the perisarc at a given time period (Fig. 3.15 B-D, dashed area). As indicated by time-lapse filming (Fig. 3.15 E), such a 'devastation' of the subperisarcal space is going on abruptly, as compared with a gradual growth of an entire stem. That confirms the idea of a non-linear triggering of a cell intercalation by stem's elongation.

2. In many cases, including the hydranth' formation some B-zone areas take during their extension a convexed shape (Fig. 3.15 D, 3.16 E, arrows; 3.21 E, pointer). This is incompatible with a suggestion of these regions to be passively pulled by more distal ones but can be easily explained as a result of the active pressure-generating extension of these areas.

3. In the Thecate stems (as well as in all the other tubular rudiments which we have traced in this respect) the curvature of a convex side is always greater than that of its opposite concave side (Fig. 3.15 A). It is hardly to imagine how such a curvatures' relation can be achieved if a stem has been bent in a passive way. Instead, it should be the natural result of SPCI to go on with a greates intensity on the convex stem's side. Accordingly, the 'devastation' of the subperisarcal space on the convex stem's side is going on with a much greater intensity than on the concave one (Fig. 3.15 B-D).

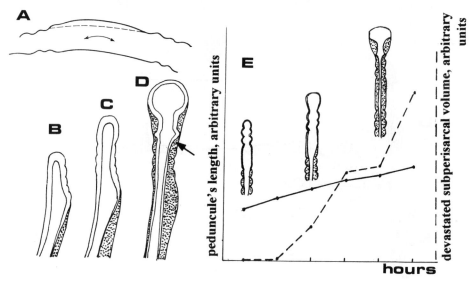

Figure 3.15. Some pictures illustrating the stretch-dependent cell intercalation in *O.loveni* development. A: the curvature of a convex (upper) peduncule's side always exceeds that of the concave side. B-D: successive stages of a peduncule's development showing the devastation of the subperisarcal space (dashed), mostly taking place on the convex (right) stem's side. E: a devastation rate plotted in arbitrary voluminal units (dotted line) as compared with the simultaneous linear growth rate of the same peduncule plotted in arbitrary linear units (solid line). Horizontal axis: hours of development. Data in B-E are taken from time-lapse film.

Another morphogenetic problem related to the branching level is: what are the initial conditions for peduncule's formation? What indeed determines a site of the origin of each new outgrowth? In the distal regions of *Obelia* colonies each new peduncule is always initiated on the convex side of a maternal one and is curved later on in the same plane. The crucial question is whether such a position is directly determined by the local geometry (namely, by the convexity), of a maternal stem or, instead, it reflects an internal asymmetric differentiation of this stem, the curvature being one of its secondary (non-causal) consequences. This alternative seems to be resolved by another Kossevitch's experiment (personal communication). The author inserted the tip of a young peduncule (destined to be bent in a definite plane) into a piece of an empty perisarcal tube, curved in another plane. The peduncule, during its subsequent growth, accurately followed the tube's curvature and, irrespective of the latter's orientation, always created the next branch on its newly convexed side. Consequently, it is the geometry rather than internal differentiation that determines the site of a new stem initiation. This may be related to at least two mechanisms, already known to us. The first of them is the 'curvature increasing rule' (see 2.3.7, pp. 141–142) based, as we know, upon the idea of a cell–cell lateral pressure. Obviously, any subsequent pressure pulse will produce evagination just on the convex, rather than the concave, side. Another cooperatively working factor is the SPCI, also increasing cell–cell tangential pressure on the convex side.

Tracing the growth of *Obelia* peduncule in more detail, we can see that its newly initiated tip is always oriented distalwards and that the stem is inclined, during the initial period of its growth, to the maternal stem (see Fig. 3.12 B, C). Both properties (which are universal for hydroids) can be explained by a proximo–distal direction of GP-associated cell-cell pressure wave and, hence, the same orientation of the SPCI–exerted pressure. On the other hand, a deviation of a daughter stem away from a maternal one, taking place during its subsequent growth period will get its explanation in the terms of a so called counterphase activity of the adjacent rudiments, to be considered later (p. 206).

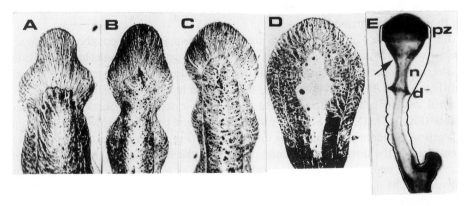

Figure 3.16. A transition from the phase of annulated growth (A, B) to the hydranth formation (C-E) in *O. loveni* development is accompanied by an extensive apical rapprochement of the both germ layers due to the distalwards endodermal sliding (see the endodermal finger-like protrusion in B, arrow). A most advanced hydranth rudiment (total view in E) has a narrow distalmost pulsatorial zone (*pz*) later on to be transformed into tentacles, more proximally situated convexed neck (*n,* arrow) and a diaphragm (*d*).

3.4.3. A LEVEL OF A PEDUNCULE'S SCULPTURING

Our previous outlook on peduncule's growth was indeed too rough for interpreting such structural details as the existence of annulations, their serial arrangement, and their alternation with the smooth peduncule's parts. Also, up to now we suggested nothing about the formation of hydranths. For doing all of this, we now have to compare the distalwards advancement rates of the adjacent ecto- and endodermal parts of peduncules. Then we can see that the peduncules, contrary to the stolons, are characterized by considerable and regular inequalities in the ectoderm/endoderm growth advances. Namely, it turns out, that during the formations of annulations the rate of the ectodermal growth advance (V_{ect}) is always greater than that of the endoderm (V_{ent}) - Fig. 3.17, cf. the plots ●---●, O----O and Δ---Δ; as a result, the annulations just formed consist mainly of the ectodermal cells (Fig. 3.16 A). Meanwhile, the initiation of a hydranth is characterized by the reverse rates relation (Fig. 3.17, same plots) and, consequently, the endodermal layer approaches much closer to the growing tip (Fig. 3.16 B-D). On the other hand, the periods of smooth peduncle's growth are characterized by an approximate equality of both germ layers' advance rates (Fig. 3.17, same plots).

Such a correlation between the growth rates' inequalities and the types of structures arised can be explained qualitatively as follows. The appearance of the annulations, initiated as purely ectodermal folds, can be regarded as a result of a loss of mechanical stability by an ectodermal sheet, pressed from below and non supported by the endoderm from inside. On the other hand, after it becomes that $V_{ent} > V_{ect}$ the situation will change drastically. Instead of being self-compressed and thus folded, the ectoderm will be passively stretched by the underlaid endoderm. That will switch on, after a certain time delay, the SPCI within the ectoderm. During hydranth development the same will lead to the formation of a thin and curved ectodermal 'neck' (Fig. 3.17 (6), 3.16 E). At the same developmental period, a close apposition of both layers in the distalmost region will promote their cooperative transverse extension. Both events are the dominating ones during the hydranth formation.

A transition from $V_{ent} > V_{ect}$ towards the reverse inequality and, possibly, back again can be derived from similar assumptions. Let us construct a morphomechanical scheme for the ecto–endoderm relations within an *Obelia* peduncule during its entire growth period (Fig. 3.17). As already mentioned, a peduncule is initiated from a columnarized ectodermal cell domain where the typical contraction–extension feedback loops should take place (Fig. 3.17 (1), (2)). SPCI seems to take place in this early growth period in the most proximal ectodermal regions, near the stem base (Fig. 3.17 (2), double line arrows). That compresses the distalmost ectoderm, leading to the formation of the 1st annulation series. As to the endoderm, it moves distalwards with a slower rate than the ectoderm and should thus be passively stretched by the latter (Fig 3.17 (2), single line arrow drawn along the longitudinal midline). At a somewhat later time, meanwhile, this very stretching will induce the SPCI within the endodermal layer (Fig. 3.17 (3)). Now the morphomechanical roles of the both germ layers are exchanged: it is the endoderm which pulls up the ectoderm, elongating a thin B-zone and providing the formation of a smooth (non-annulated) peduncule's region. Within some time period, the next reversion takes place: the SPCI is initiated again within the ectoderm, bringing into being the formation of the 2nd annulation series (Fig. 3.17 (4)). The next round of similar passive/active extension swings will take part during the formation of a hydranth. At its beginning, the endoderm becomes active again, exceeds the ectodermal distalwards advancement rate, and creates thus a hydranth primordium (Fig. 3.17 (5); 3.16 C, D). Then the stretched ectoderm takes the active role for the last time and moulds, as a result of SPCI, a thinned 'neck' of the hydranth (Fig. 3.17 (6); Fig. 3.16 E). It is obviously an almost complete exhaustion of the

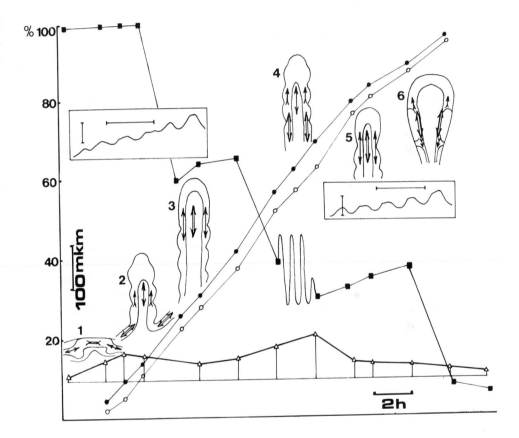

Figure 3.17. A combined plot and a morphomechanical interpretation of the processes related to the sculpturing and branching levels of *O.loveni* peduncule's growth. 1-6: successive developmental stages together with the suggested stress patterns (those assumed to take place in the endoderm are plotted along the midlines). Horizontal axis: time (hours). Vertical axis: percents of a peduncule's length occupied by the pulsatorial (A-) zone at the different developmental stages (as related to the ■---■ plot) and a total peduncule's length, see vertical bar (as related to the ●---● and O--O plots). A wavy region in ■---■ illustrates considerable fluctuations in A-zone length taking place at stage 4. ●---● and O--O show the coordinates of the distalmost points of the ecto- and endoderm, respectively. Δ---Δ gives the differences between these points coordinates (that is, the thickness of a tip ectoderm); linear scale is twicely enlarged as related to other graphs. Framed are GP patterns typical for two developmental periods: stage 1 (left box) and a transition between stages 4 and 5 (right box). Within the boxes horizontal bars are 10 min and vertical bars are 10 mkm.

pulsatorial A-zone which brings these passive–active extension swings to their end. Is it possible to trace any elements of a *hyper*restoration within these swings? At least two such events can be pointed out. Firstly, at the GP level one can trace, within certain developmental periods at least, a gradual increase in the amplitude of each successive pulsation as compared with the preceeding one (Fig. 3.17, framed GP records). Secondly, in the realm of the more prolonged T_{ch}-s it is the amplitude of the second ectodermal growth advance which is definitely greater than the first (Fig. 3.17, plot Δ---Δ).

Our general conclusion would be that the antiphase relations between the activity/passivity of both layers' stretching and, hence, between their SPCI reactions are the main conditions for the differentiation of a peduncule. It is also obvious that all of these processes can be regulated, in a most effective way, by the rate-determining parameters, to which, as already mentioned (see Chapter 1), the genetical factors as well as a number of epigenetical ones can be attributed. As a kind of a rough approximation we shall qualitatively estimate how a stem's sculpture can be modified by the modulations of only two out of all the possible temporal variables, namely:

(1) by the duration of a stretch–intercalation delay;

(2) by the rate of a cell intercalation traced separately within each germ layer.

We shall denote a stretch–intercalation delay in the ectoderm as SID_{ect}, the same in endoderm as SID_{end}, the rate of intercalation in the ectoderm as IR_{ect} and the same in endoderm as IR_{end}. It is obvious, first of all, that if $SID_{ect} \approx SID_{end} \approx 0$ a stem will grow in a smooth, stolon like manner without producing any macroscopic morphological structures (Fig. 3.18 A). If now SID_{ect} goes up while SID_{end} remains negligibly small, the endoderm will continue to elongate itself in a smooth manner, while in the ectoderm the thinned and thickened zones will alternate each other, the greater be SID_{ect}, the larger become these zones (Fig. 3.18 B). Generally, under these conditions the annulated pattern typical for the peduncules rather than stolons will be produced (Fig. 3.18 B, C). Its characters will depend meanwhile upon the properties of the endodermal growth advancement, that is, upon the relative values of IR_{end} and SID_{end}. For example, a high IR_{end} value will diminish the annulations' thickness and promote their uniform spreading throughout prolonged peduncule's distances (Fig. 3.18 C). Meanwhile, a great enough SID_{end} value will produce the periodic thickenings of endoderm as well (Fig. 3.18 D). When both SID_{end} and IR_{end} will exceed the similar ectodermal's values these periodic thickenings will be even more pronounced and covered with just a thin layer of ectoderm (Fig. 3.18 E). In this way we can approach, although up to now in a very rough approximation, the morphogenetic interpretation of the different branching types in hydroid polypes. Meanwhile, their more subtle species-specific morphological characters as well as those related to the patterns of stem bendings (shown in Fig. 3.18 F, G) are associated with the activities of a lower level, namely that of the growth pulsations and will be thus commented in the next section.

Fig. 3.18 (see opposite page) . Architectonical properties of Thecate outgrowthes which can be attributed to the differences in the stretch-intercalation delays, rates of intercalation and the "width" of growth pulsations. As before, double-line arrows indicate the suggested active stresses and the apicalwards shifts of the cell material while one-line arrows indicate the same characters established in a mechanically passive way. For more comments see text.

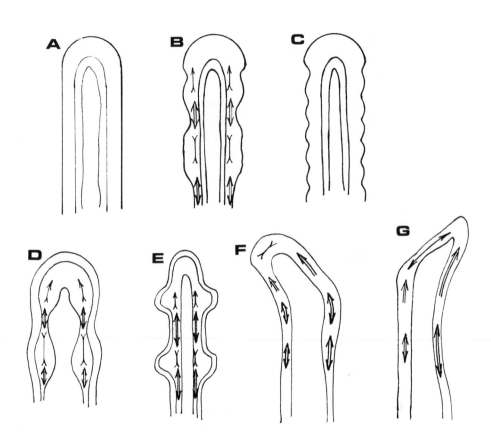

3.4.4. A GROWTH PULSATIONS (GP) LEVEL

Direct morphogenetic effects of GPs.
Different species of hydroids and even the different developmental stages of one species
exhibit rather different GP patterns, which can be roughly estimated by the values of the
normalized integers of the GP records curves (Fig. 3.19 A–E). These patterns are
precisely regulated, both at genetical and at epigenetical level (Wyttenbach et al.,
1973; Beloussov et al., 1985). According to their GP patterns, the different species of
hydroids can be roughly divided into those exhibiting 'narrow', 'intermediate' and 'wide'
GPs. The 'wider' is each successive GP, the more extended in the transversal plane
should be the hydranth rudiment (Fig. 3.19, cf A and B; C, D and E). The reason is
largely mechanistic, and may be illustrated by the practice of glass blowing: the

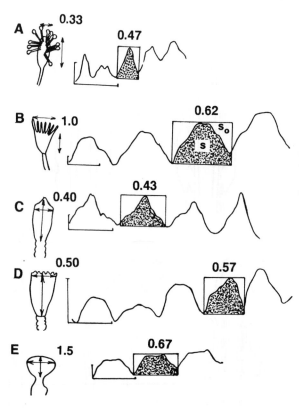

Figure 3.19. Correlations between the species-specific hydrothecal shapes (width-length relations, see values at left) and GP patterns, as measured by S/S_o relations. S is a time-extension integer (a square under the curve recording tip shifts) and S_o is the period-amplitude product for the same GP. Two Athecata (*A, B*) and three Thecata species (*C-E*) are presented. *A: Coryne loveni, B: Eudendrium rameum C: Campanulina lacerata, D: Obelia loveni, E: Dynamena pumila.*

specialists know that in order to produce narrow elongated vessels the pressure pulses should be brief, while for making a wide vial more prolonged blowing efforts are required. Just the same kind of work is done by the pulsating hydroids tips which mould, by the pulses of cell–cell pressure, a still soft distalmost perisarc. More accurately, the transverse extension of a sample requires greater efforts than the longitudinal extension because, as we know, the circular stresses are greater than the longitudinal ones. The next point which is worth mentioning is that a force causing the transversal extension after passing a certain threshold, well may break the radial symmetry of an extended rudiment exchanging it by a reduced order symmetry (most probably, by a mirror symmetry). A gradual transition from the initial radial symmetry of the transversal sections towards the mirror symmetry is quite obvious, for example, in the colonies of a hydroid, *Dynamena pumila,* to be described later one in greater detail.

Indirect (secondary) morphogenetic effects of GPs.
GP-mediated changes in branching types. Besides the above described primary effects, the GP-mediated shape changes can lead to some important secondary consequences. Most of them are associated with the abovementioned reductions of a growing tip's symmetry order under the influence of strong enough (that is, 'wide') GPs. Suggest, for example, that a radially symmetrical tip is transformed in this way to a mirror symmetrical one. According to the curvature increasing rule (see 2.3.7, 3.4.2) such a rudiment's roof will be separated, along the direction of its maximal extension, into three parts by two vertical indentations. These indentations may become the sites of the new pulsatorial zones located now much more apically than in *Obelia* samples. In such a way a so-called apical, or a monopodial, branching type is established, characterized by a prolonged growth and budding activity of a single non-exchangeable tip. This is opposed by a so called subapical, or a sympodial, type where a new branch is initiated each time from a new site, localized much more towards proximal, like in *Obelia*. Accordingly, a sympodial type of branching is correlated with the narrow growing tips, and with more narrow GPs.

Let us consider now the asymmetric stems, that is, those with one of their sides extending in a proximo-distal direction with a greater rate than the opposite (the origination of such an asymmetry will be interpreted in the next section). This may have two consequences, depending upon a stem's tip extensibility (and, hence, in the long run, upon GP 'width'). If a tip remains narrow and, correspondingly, a stem's diameter remains constant, such an asymmetry will lead, as in *Obelia* samples, to a gradual stem bending: its more extended side will become convex, while the opposite one becomes concave (see Fig. 3.18 F). If, however, a stem tip can be extended transversely (either passively or actively) thus taking the shape of an oblique roof, the mostly elongated side of the stem will take, because of the active internal pressure and a roof's mechanical reaction, a concave shape, while the opposite one remains more or less straight (Fig. 3.18 G). Our final conclusion will be that the GP pattern should affect at least three mutually independent macromorphological events: the transverse extension of a tip, mode of branching, and the overall shape (curvature) of a stem. For example, the 'wide' GPs provide, at the same time, transversely extended tips, monopodial branching and straight stems, while the 'narrow' GPs produce narrow tips, more curved stems, and a sympodial branching type. If our assumption is true, no one member from the first triple can be combined, within the same species, with any one from the second triple, and vice versa. So far as we know, this prediction has no exceptions at least within the Thecates.

Going into some more detail of the branching patterns we can distinguish within a monopodial type itself three different modifications: a so called unilateral, an asymmetric (alternate), and a bilaterally symmetrical (opposite) branching. The passage from the first towards the third can be regarded as the result of a more or less correlated increase of an entire set of the parameters providing the transversal extension of a growing tip (and in particular those increasing GP width and IR_{end}): slightly extended but asymmetrical tips should exhibit unilateral branching, more extended ones will be branched in the alternate way and most of all extended will show the opposite branching. Such is just a temporal succession of the branching patterns in the juvenile colonies of *Dynamena pumila*, developed directly from the settled planula larvae.

Counterphase effects. While tracing, by time lapse filming, the patterns of the growth pulsations within the lateral indentations of the tripartite *D. pumila* tips we have observed a remarkable 'counterphase effect' (Beloussov, 1991): the cells situated on the opposite sides of the same indentation, were pulsating in counterphase to each other. Namely, when the cells belonging to the lateral rudiment took the transverse orientation (corresponding, as we know, to the extension GP phase), those belonging to the central rudiment appeared to be in oblique positions (corresponding to the retraction GP phase), and vice versa (Fig. 3.20). Obviously this is the most rational regime, providing mutual aid rather

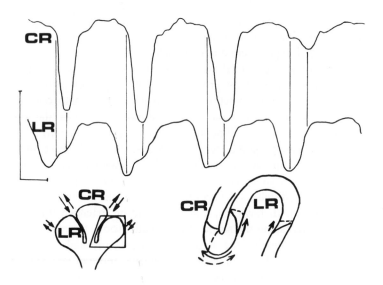

Figure 3.20. Counterphase GP activity in the adjacent walls of the lateral (LR) and central (CR) rudiments of *Dynamena pumila* growing tips. Upper row: a simultaneous record of the periodical shifts of the tips in the adjacent CR and LR. Vertical lines indicate coincidences of CR retractions with LR extensions Lower row: schemes of the counterphase movements (from Beloussov, 1991).

than the mutual prevention of cell rotations in the opposite walls of each indentation: the same is done by two persons piling logs. The result will be not only a facilitation of the cell shifts but also an increase in their amplitudes. As a result, in the case of counterphase activity the cells of the adjacent walls of a central and a lateral rudiment (separated from each other by an indentation) will move distalwards at a greater rate than those of the external walls of the same lateral rudiments which have no counterphase partner; as a result, the lateral rudiment should be deviated more towards lateral, what is just the case for *D. pumila* (Fig. 3.21 A). A crucial experiment would be to check, whether the lateral rudiments will grow more vertically if a central one will be excised, or whether the central rudiment, after removal of the lateral one, will be inclined towards the excised rudiment. These experiments have been indeed carried out on *Dynamena pumila* samples (Beloussov, 1973), and in the both cases the expected results have been obtained - Fig. 3.21 B-E. It is worth mentioning that after the removal of a central rudiment the lateral

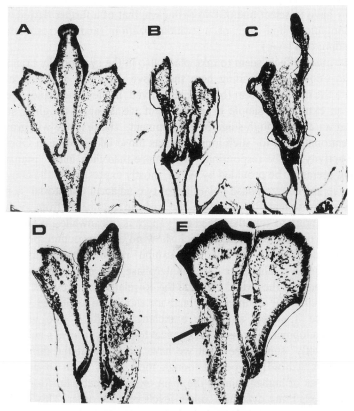

Figure 3.21. Rudiments' interactions within the apical whorl of *Dynamena pumila* colonies. A: intact whorl. The central rudiment (CR) is perfectly symmetrical while the walls of the lateral rudiments (LR) are equally deviated from CR. B, C: as a result of an early CR removal, the both LR-s take the vertical (non-deviating) positions and elongate themselves with the different rates. D: after the removal of the right LR, CR becomes inclined towards right. E: a similar operation shown at the earlier time moment, left LR is removed. On the left side of CR a pronounced cell fan appears (arrow), indicating the active contractile response of the understretched left CR's wall.

ones not only change their orientation, that is, start to grow more vertically but show also much less precision both in their shape and growth rate (Fig. 3.21, cf. A and B, C). Similarly to what have been pointed out for amphibian embryos we can see that a disruption of a holistic mechanics affects a preciseness of the morphogenetic processes to a much greater extent than a developmental dynamics itself. Some of the experiments on *D. pumila* whorls gave also clear evidences that the contraction-extension feedback (2.5.1) works perfectly within the hydroids' epithelia as well. As seen from Fig. 3.21 E, after the removal of the left lateral rudiment the adjacent wall of the central one becomes understretched. Its only way for restoring the normal tension will now be cell columnarization. We can see that this is just the case and that the perfect 'fan' created by highly columnarized cells is formed indeed in the corresponding part of the ectodermal layer (arrow). Thus the hydroids' cells seem to obey a bistability regime to no less an extent than those of amphibian embryos: as depending upon the amount of a layer's stretching, they have to select one of two pathways, that of a further flattening (and hence rudiments' elongation) and that of a columnarization (and, correspondingly, of a subsequent folding of a layer).

The counterphase effects seem to take place also in the sympodially branching species, such as *Obelia* colonies, providing here the above described deviation of a daughter peduncule from the maternal stem (both being curved in the same plane). Such a view is supported by an extremely simple experiment: if we isolate a daughter stem from the maternal one at an early enough stage of development, the first one will grow straightly. It is worth mentioning that for such long stems as those taking place in Obelia colonies a counterphase activity can be responsible for no more than their initial mutual deviations: further bending seems to be provided by SPCI, mostly expressed on the convex side of a stem. In any case, the counterphase relations of two adjacent pulsatorial zones look like being the main instrument for deviating the hydroid rudiments from each other, thus providing better conditions for their feeding. There is, meanwhile, a number of species (by all estimations, no less flourishing) which show no obvious counterphase activity (and hence no stems deviations) even if they exhibit monopodial branching and apical indentations (see below). Therefore the counterphase activity cannot be regarded as a necessary consequence of the 'wide' GPs and the correlated set of events.

Generally speaking, the counterphase activity should be attributed to the branching level's events since it affects an entire branching period. A role which it plays in maintaining the developmental order is also typical for an upper level's activity.

Let us now briefly summarize what we have learnt about the different structurally dynamical levels in hydroid polypes. Although their levels seem to be far from being homologous to those of higher animals the main morphogenetic functions are very much the same: lower levels bring dynamics, upper levels provide order. This is especially clear if comparing poorly organized stolons (which do not have the levels situated immediately above that of GPs) with much more accurately organized peduncules, possessing a complicated multilevel hierarchy. (Another thing is that a poor stolonial organization has its own advantages as providing possibilities for a random search!). The way used by hydroids in creating their upper levels is also worth to be commented. The task may be formulated as follows: to construct a higher level (that characterized by slower rates, that is, by greater T_{ch}-s) directly out of the lower level's processes (that is, more rapid ones). A solution is really elegant: it is in using the difference between the rates of the two lower levels variables, namely V_{ect} and V_{ent}, as a basic measure for a higher level's variable rate. The difference is small and so is, hence, the upper level's variable rate! Somebody: God, Evolution or Hydrozoa themselves behaved as witty mathematicians (although probably a bit too linearistic)!

It is, meanwhile, of interest that such a linearistic way for making the upper levels from the lower levels' components did not retained as the main one during a subsequent evolution of the animal's kingdom: as we could already see, the interlevels relations in the vertebrate embryos in no way can be reduced to such a simple scheme. Also, in higher animals the levels are much more removed from each other according to their T_{ch}-s and L_{ch}-s values than those traced in hydroid polypes.

3.4.5. FROM A COMMON MORPHOGENETIC ARCHETYPE TO THE DIVERSITY OF THE SPECIES-SPECIFIC SHAPES IN THECATES

Let us try now to give a concise description of the Thecates development by constructing a kind of their 'developmental archetype', namely a succession of the morphomechanic events which will be common for all of this group representatives and will consist of units derivable from each other on the basis of the known rules. As an initial stage of such a reconstruction we would like to take that one most of all simply organized. Such is, for example, a piece of a stolon. Normally the Thecates only rarely commence their development from this stage exactly, preferring to start it from a more structured formation (say, a growing tip of a colony) for using in its subsequent development some of this previously established organization. Meanwhile, the latter cases can be also easily inserted into the same general framework.

The successive steps of a suggested developmental archetype are as follows:

(a) The appearance of a new pulsatorial zone on the surface of a stolon piece. As said before, this is associated with the formation of a columnarized cell domain and can thus be treated as a typical self-organizing phenomenon (see p. 125, Fig. 2. 25).

(b) Establishment of asymmetry and of the initiation sites of new stems. A stem's rudiment, if it does not interact (in a counterphase manner) with another actively growing rudiment, should remain roughly symmetrical (straight). Such a symmetry is, however, unstable, the type of instability being different for the stolons and for the peduncules. In the first ones the instability can be qualified as indifferent (non-robust), insofar the locations and the amplitudes of the growing asymmetries ultimately depend upon the external factors. As to the peduncules, the asymmetry is regulated by the adjacent rudiments and the essentially robust mechanisms of asymmetry reinforcement (mostly associated with SPCI) are coming into action. As depending upon GP patterns, they lead, as we know, either to a smooth bending of a peduncule or to the formation of an oblique extended roof covering the peduncule which remains almost straight, although if becoming asymmetrical (see Fig. 3.18 G, H). In both cases new rudiments will be initiated on the most extended side of a maternal stem.

(c) Regular deviations of the neighboring rudiments from each other owed to their counterphase activity. Sooner or later at least two interacting (mechanically bound) pulsatorial zones should appear. As shown above, they start to work in a counterphase with each other, providing thus a mutual deviation (as well as a structural asymmetry) of both rudiments. The asymmetry/deviation thus becomes strictly regulated.

(d) Detalization of a stem's structure owed to temporal shifts in the ecto/endodermal growth advancements. As mentioned before, the passive stretching and the active SPCI within the ectoderm is going on, as a rule, in the antiphase with the same processes in the adjacent part of the endoderm, an vice versa. The antiphase activities may be for several times reversed, providing thus the formation of periodic structures, for example the annulation series. These relatively frequent are combined with a more long termed tendency for a gradual diminishing of the apicalmost pulsatorial zone. This tendency, together with the endoderm-dependent transversal extension of a growing tip leads to the

termination of a peduncule's growth and the formation of a single hydranth (as in *Obelia*) or an entire apical whorl (as in *Dynamena*).

All of these events, with the specific exception of the counterphase activity are based upon several well known morphogenetic reactions: contact cell polarization, tendency for a local curvature increase, and stretch induced cell intercalation. Each of them can be regarded as the manifestations of a universal robust tendency for a (hyper)restoration of the mechanical stresses. The Thecate hydroids were probably among the first in the Metazoans' evolution which have learned to use this (initially adaptive) tendency for morphogenetic purposes, and this was never lost during a subsequent evolution of the animal's kingdom. It is also worth to emphasize that the mentioned events are may be coupled with each other only *via the macromorphological properties* of the samples, such as the curvatures, stem's sculpturing, topology of pulsatorial zones, etc.. Without these intermediate macromorphologies, no regular morphomechanical successions could take place at all.

Our next task will be to explore, at least qualitatively, to what extent the main species-specific differences in the Thecate shapes can be considered as the modulations of just these morphogenetic reactions, assuming that their rate determining parameters take different values in the different species. In the other words, we shall try to create, in most general outline, a morphogenetically oriented taxonomy of the Thecates. Ideally, it should take the form of a multidimensional graph, each of its dimensions corresponding to the changes of the the certain parameters' values. As usually, we shall classify the parameters according to their T_{ch} -s. Within such a framework the following categories of the parameters can be distinguished:

(a) Those modulating the GP-associated T_{ch} -s and thus affecting directly the 'width' of the GPs. We shall plot this set of parameters along the X-axis and call them X-parameters.

(b) Those modulating the processes with a the greater T_{ch} -s, namely those responsible for the peduncule's sculpturing and branching pattern. This is, as we know, a vast and heterogeneous group of parameters, but now we would like to regard all of them from a single point of view: namely, how do they affect the V_{end}/V_{ect} ratio value. We will plot these parameters along the Y-axis, starting from those providing the lowest V_{end}/V_{ect} values (and hence producing long annulated stems with rarely arranged hydranths) up to those corresponding to the highest values of the same ratio and consequently to the short and transversely extended non-annulated stems, settled by densely packed hydranths. This will be the Y-parameters group.

(c) Finally, we shall pass to the parameters associated with even greater T_{ch} -s values; they can be hence roughly considered as temporally constant ones. Within this group we shall outline, in particular, the parameters affecting in some way the mechanical properties of hydroid tissues and the rate of perisarc deposition. All of these mechanical parameters we will unify under the notion of a *generalized elasticity* of the germ layers, that is their tendency to restore, as a result of the elastic forces action, the initial layer's curvature after its active changements (see p. 143, Fig. 2.34 and the corresponding comments). The greater is the generalized elasticity, the smoother should be the rudiments' contours, and vice versa. We will not use a special axis for plotting these parameters because we suggest that in each small taxonomic group (no greater than of a genus' range) they can vary independently from all the other parameters. Meanwhile, we shall pay a special attention to another related group of mechanical parameters which are capable to modulate the counterphase activity, most probably by affecting the elastic properties of those parts of the mesoglea which connect two pulsatorial zones . Obviously, too soft and non-elastic mesoglea will hamper counterphase interactions, and vice versa. This group of parameters will be plotted along the Z-axis and called the Z-parameters group.

We will arrange most of the Thecates within the X–Y plane (Fig. 3.22), only for a few of them using, as an addendum, the Y–Z plane (Fig. 3.23). Let us start our travel along the X–Y plane from its left lowest corner (Fig. 3.22). Firstly, we shall move horizontally to the right at a low Y-level. In this way we will rest ourselves within the realm of the species with long annulated non-branched stems with rarely packed hydranths. We start

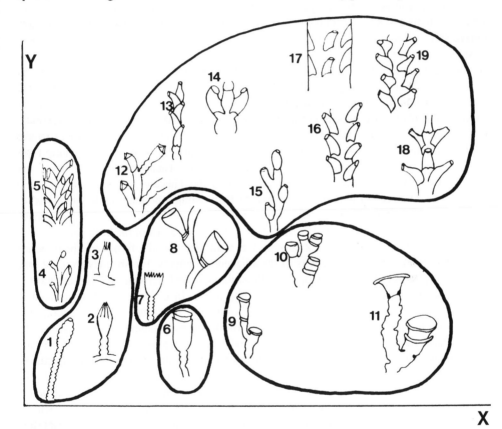

Figure 3.22. The Thecate taxonomy in X, Y parameters space. Families are contoured. 1: *Eirene indicans*, 2: *Opercularella nana*,   3: *Lafoeina tenius*, (1-3: family Campanulinidae). 4: *Cryptolaria flabellum*, 5: *Grammaria immersa*, (4, 5: family Lafoeidae). 6: *Bonneviella regia* (family Bonneviellidae). 7: *Campanularia johnstoni*, 8: *Obelia loveni* (7,8: family Campanulariidae). 9: *Halecium mirabile*, 10: *H. beringi*,   11: *H. corrugatum*   (9-11: family Haleciidae). 12: *Thyroscyphoides biformis*, 13: *Hydrallmania falcata*, 14: *Abietinaria labrata*, 15: *Sertularella polyzonias*, 16: *Thuiaria articulata*, 17: *Thuiaria pinna*, 18:*Sertularia staurotheca*, 19: *S. mirabilis* (12-19: family Sertulariidae).

from some typical Campanulinidae (such as *Eirene* and *Opercularella*), then go via *Campanularia* genus and Bonneviellidae family, and come at last to Haleciidae with the widest, often trumpet like, hydrothecae, produced obviously by 'wide' GPs. A vertical upwards movement in the immediate vicinity of the Y-axis will bring us firstly from the abovementioned Campanulinidae with prolonged annulated stems to those with progressively reduced stems and tubular hydranths (such as *Cryptolaria*) up to the fully developed Lafoeidae (*Grammaria*) with their densely packed (corresponding to high enough Y-parameters' values) and, at the same time, narrow tubular hydrothecae (as derived from low values of the X-parameters).

The mostly populated part of the table, probably reflecting the main evolution trend within Thecates is its diagonal, corresponding to concerted increase of the both X- and Y-parameters values. Here we pass, at first, from the Campanulinidae with narrow annulated stems and a very rudimentary branching to the *Obelia* representatives with more transversely extended hydranths, well developed sympodial branching and stems possessing non-annulated zones. Both latter events point indeed to an increase of the endodermal growth activity, as related to the ectodermal one, that is, to the increase of Y-parameters values. *Obelia* genus, by all estimates, occupies a central position in the Thecates' taxonomy, so that from this point we can move not only further along the diagonal, but also more or less vertically upwards (thus making the colonies ever more 'endodermized'). In this way we reach (deviating probably a bit to the left that is, towards smaller X-values) the *Thyroscyphoides* genus, still keeping a sympodial branching type, and, later on, the unilaterally branched *Hydrallmania*. These two little genera are intermediate between the two largest families among the Thecates, Campanulariidae and Sertulariidae. We reach the latter by moving further along the main table's diagonal to the upper right. The densely packed hydranths, which are at the same time very much deviated to the lateral, correspond to the highest possible values of the both parameters.

As to the parameters from Z-group, we shall use mostly those of them which characterize the counterphase activity of the adjacent rudiments. We shall try to regard the specific morphological characters of some Plumulariidae and Sertulariidae representatives as the functions of these parameters values (Fig. 3.23). In order to do this we have to remind that the counterphase activity can be estimated by comparing the lengths and curvatures of the external and internal walls of the lateral rudiments which are in the counterphase relations with the adjacent central ones. Obviously, in the cases of high counterphase activity the internal walls of the lateral rudiments (the only ones objected to the counterphase activity) should be more extended than the external ones, so that the former will be definitely convex, while the latter will be straight or concave. Meanwhile, with the decrease of the counterphase activity the lengths of the opposite walls will be firstly equalized (so that the branches or the lateral hydranths will become straight) and then may be curved to another side, that is, the external walls will become convex.

Now we may conclude that the representatives of the Plumulariidae family are distinguished from the most other Thecates by a very low level of counterphase activity, as indicated by a condensed arrangement of their numerous appendages and by the convexity of their external walls (Fig. 3.23, 3). On the other hand, as we have mentioned before, the low counterphase activity correlates with a relatively low tension of the mesoglea. This very peculiarity may promote extensive and detailed sculpturing so typical

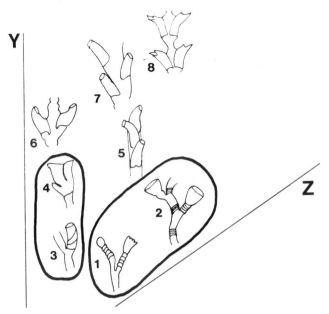

Figure 3.23. An appedix of the Thecate taxonomy plotted in Y, Z parameters' space. 1:*Obelia loveni*, 2:*O. geniculata* (1,2: family *Campanulariidae*), 3: *Plumularia linkoi*, 4: *Nuditheca dalli* (3,4: family *Plumulariidae*), 5: *Hydrallmania falcata*, 6: *Abietinaria labrata*, 7: *Thuiaria involuta*, 8: *Dynamena pumila* (5-8: family *Sertulariidae*). All the taxonomic diagnoses used in Figs 3.22 and 3.23 are taken from Naumov (1960).

of this family's representatives. At the same time, a complete lack of the annulated zones and rather dense arrangement of the hydranths indicates that these species' Y-parameters values should be estimated as rather high. On the other hand, they are considerably extended along Y-axis, because the different Plumulariidae show different degrees of 'endodermization' (compare *Plumularia linkoi* and *Nuditheca dalli*, Fig. 3.23, 3 and 4). Coming back to the *Obelia* species, we may suggest that the counterphase activity should be increased while moving from *O.loveni* with its more condensed stems towards *O.geniculata* with its much more deviated branches. And, finally, within the Sertulariidae family we should arrange along Z-axis from left to right firstly the species of *Abietinaria* genus (as characterized by the convexity of the external hydranth's walls), then the *Thuiaria* representatives with their hydranths walls almost straight and, lastly, such species as *Dynamena* with the lateral hydranths possessing convex internal walls and considerably deviated from the central axis of a branch.

The suggested table requires some special comments and reservations. Firstly, it should be noted that even in the best case it would be unrealistic to reach, for the biological species, the same kind of unambiguity between the position within the table and the sample's properties as, say, for the chemical elements in the periodical table. The first, and probably the main, reason of the ambiguidity is that the phylogenetic history of a species always affects its mode of ontogenesis, even if not to the extent declared long ago by Haeckel. Consequently, the ontogenesis, including shaping, is also a function of a species' phylogenetic history, which plays a role of initial conditions rather than permanently acting developmental parameters. Returning to Thecates, we can speculate that their main evolutionary trend roughly corresponds the table's diagonal, going from the lower left to the upper right. Correspondingly, the evolutionarily most advanced species (shifted to the upper right) should be to a greater extent loaded by the evolutionary 'reminiscences'. This can be seen, for example, in *Dynamena* colonies which, while developing from the just settled planulae, exhibit firstly the Campanulariidae–like radial symmetry, then the *Hydrallmania*–like unilateral branching, only later coming to their specific opposite branching pattern.

Another side of the coin is that the different kinds of parameters can interact with each other, being able to produce some unexpected non-additive effects, either promoting or preventing the appearing of new structural details. This is why the different parts of the table are populated rather unevenly. To this we may add that the parameters themselves are also interdependent: as mentioned before, it is the increase of the both X- and Y-parameters' values which brings into being a new Z-parameter, associated with the counterphase activity. Also, the parameter's values may be changed from one developmental stage to another.

A specific point which deserves to be discussed is how might one pass from the global architectonics of a species-specific shape to much more local properties, commonly used by zoologists for the taxonomic purposes. Is it possible to consider at least some of the latter as the direct functions of the same developmental parameters? Let us explore such a possibility, taking as an example a ubiquitous taxonomic character of this group, namely the presence and the site of attachment of an operculum, covering a hydrotheca opening in some Thecates. What might be, first of all, the global morphogenetical conditions for the presence of an operculum?

They are very simple indeed: in order to have an operculum, a hydrotheca's opening should not be too wide. That puts automatically the operculum-bearing species at two extremal locations within the main table (Fig. 3.22): firstly, in the left lower corner, which is occupied by the species with solitary narrow hydranths, and secondly in its right upper section populated by the whorled species whose lateral hydranths are again narrow. There is, meanwhile, a basic architectonic difference between both mentioned groups of the operculum bearing hydranths: while the first are radially symmetrical, the latter are asymmetrical. This affects the structure and the mode of the operculum attachment: in the first group it is radially symmetric, consisting of several equal petals, while in the second it is itself asymmetric. Then we can see that in the whorled species the operculum is attached to that side of a lateral hydranth which is growing slower than the opposite side (obviously because a slowly growing wall is the most suitable place for the deposition of a perisarc, creating the fixed side of operculum). Consequently, the place of the operculum's attachement appears to be directly linked with the counterphase activity: in the species with a low counterphase activity (such as *Abietinaria*) it is attached to the internal side of a hydranth while in *Thuiaria* species expected to have a greater counterphase activity it is attached to the external side.

Another widely used taxonomic criterion is the indentation of the hydrotheca opening: some of Thecate species have sharply indented hydrothecae while another closely related species exhibit smooth openings. A reduction of this extremely local character to the global morphogenetic parameters represents an intriguing and as yet unsolved morphomechanical task. One may suggest that the indentation depends at least upon two categories of parameters: the general elasticity tending to smooth out any local curvature differences (and hence indentations) and the rate of perisarc deposition (the greater the deposition rate the more pronounced should be the indentations). A high rate of a perisarc deposition can also be responsible for the appearance of the densely arranged circular ribs on the surface of some species hydrothecae (such as *Sertularella tenella* Alder or *S. sinensis* Jaderh). In these cases literally each next GP is obviously 'frozen' by the rapidly deposited perisarc.

We see no principal obstacles for deriving, by following the same way, all of the taxonomically useful local characters from the global parameters' values: this will really bring a classical taxonomy and a developmental biology much closer to each other. And insofar the parameters employed are opened for a direct genetical determination (as containing nothing more than the rates of the processes) our approach can be also regarded as a genetically oriented one. Meanwhile, its relations to the conventional genetics are rather specific and deserve special comments.

3.5. On morphocentric and genocentric approaches: harsh opposition or a hope for fruitful collaboration?

The approach pursuited throughout this whole book (and expressed most overtly in the last Chapter) can be called morphocentric insofar its main aim is in suggesting some fundamental laws of morphogenesis which cannot be immediately reduced to the notions of genetics. On the other hand, we always tried to formulate these laws in a way permitting them to be parametrically regulated by the factors which can belong to the immediate products of the genes' activity. In such a way we are trying to pass our part of a way from morphogenesis towards genetics, although it remains quite far from our aims to identify the individual genes involved in the regulation of a given developmental process. An alternative, more ubiquitous and fashionable modern approach may be characterized as a genocentric one since it starts just from the opposite point, that is, from the genes. By its extreme (but, nevertheless, widely accepted) version the more will we succeed in discovering the genes which are expressed during a given developmental process and which affect it in some way the better will we understand this process. At the first glance, the both approaches look as being mutually symmetric and complementary, having thus a good chance to meet each other at a certain midpoint and to reach, at this moment, a full description, or an explanation, of a developmental process. Actually, however, the situation is far from being so idyllic. The matter is that the epistemological structure of the both approaches is quite different. Whereas a morphocentric approach is a derivate of a nomothetic ideology directed towards establishing some invariable embracing laws of a process under study, the genocentric approach can be qualified, may be crudely, as a 'button-like' one. A belief that the identification of all the genes involved into a given

developmental   process   will   provide us   by   its   full description or explanation very
much resembles the conviction of a person who have learned the functions of the buttons
on a TV panel (knowing at the same time nothing about electromagnetic fields, etc.) that
he or she gets now a real 'understanding' on how the device is working out. Actually, as
applied to a developing organism such a "button-like" ideology is even less adequate than
if being related to a man–made device: suggest that we have a TV or a radio set which
issues different programms after switching on the same button and vice versa or which,
even worse, starts the buttons by its own willing without requiring at all your intervention.
And this is just the way used by a developing embryo in operating with its genes. We
know perfectly that the latter ones not only are switched on or off during the definite
developmental periods and at the definite sites   (as if obeying a certain epigenetic
programm) but, in addition, that the same sets of regulatorial genes become involved into
the regulation of quite different developmental processes, such as e.g. gastrulation,
neurulation or limb formation (see, e.g., Shubin et al., 1997). Similarly, "the interactions
between msx-1 an msx-2 homeodomain proteins characterize the formation of teeth in the
jaw field, the progress zone in the limb field, and the neural retina in the eye" (Gilbert *et
al.*, 1996). We know also, that in many cases no one-to-one correspondence can be traced
between the patterns of genes expression and the morphological structure of the
corresponding organs. For example, a recent study of the genetical expression
compartments in the embryonic rudiment of a *Drosophila* leg showed that some of its
single segments (tibia) include more than one such compartment while in other regions
(tarsal segments and claws) several segments exhibit the same expression pattern (Lecuit
and Cohen, 1997). A similar example related to *Arabidopsis* plant is given by Goodwin
(1994, pp. 126-128): a gene responsible for the number of flower's petals is
homogeneously expressed throughout all of the petals' whorl. Even greater is the
incompatibility between the immediate products of the genes activity and the genetically
affected morphogenetic movements. All the morphological events, let they be regarded in
ontogenetical or in evolutionary aspects (as D'Arcy Thompson (1942) did by using his
famous coordinates' transformations) are markedly holistic and thus irreducible to any
microscopical scalar heterogeneities, the only possible immediate results of a genetical
regulation. So, the genes look much more like obedient servants summoned up from time
to time in order to fulfil a required work (even without knowing its results in all the
details) rather than a supreme power having its goal within itself.

How such a situation can be interpreted in more or less precise terms? A general
answer is already known: the genoproducts play a role of parametrizers of the
developmental feedback contours. So, paraphrasing what have been already told (p. 167)
we may claim that to succeed in identifying the individual genes without knowing, even in
general outlines, these contours *per se* is the same as to know the parameters' values
without having any idea about the equation(s) in which they are figured out: that means
to know nothing. On the other hand, a knowledge of the equation (or of a feedback
circuit) *per se* even without specifying the parameters' values is enough, as we know,
for getting some general image of a system's behaviour (at least, in the terms of its
potencies). But certainly to know, in addition, the parameters' values would be even
better. If we accept, at least in its general outlines, such a point of view, the interactions

between the morphocentrical and genocentrical approaches will become more balanced and heurisctically fruitful than they are now.

Meanwhile, even if accepting the idea of a parametrizing function of the genome we become confronted with the fact that these parameters are, in a majority of cases, rather unusual. Only rarely can they be treated as 'canonical' parameters which must preserve constant values throughout an entire space–time of the development. One of such rare cases is exemplified by a so called allometric growth (Huxley, 1932) which is known to be characterized by the constancy of the logarithmic ratios of the growth advancement of a given organ in two different dimensions. In several plants' species the coefficient of the allometric growth (that is, the relation of the logarithmic ratios) is known indeed to be directly determined by the genes which probably affect the mechanical properties of the cellulose cell walls (Sinnot, 1960). This coefficient takes very precise and constant values within vast enough embryonic regions and prolonged developmental periods (Fig. 3.24). Thus, it may be really attributed to the 'canonical' parameters.

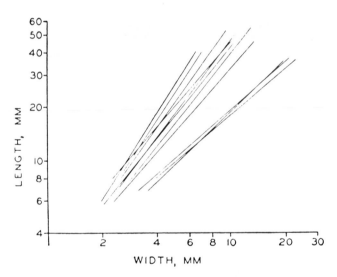

Figure 3.24. Segregation of allometric growth rates in F_2 from a cross between a rather elongate and a rather flat variety of cucurbits. The two F_2 classes resemble in general the parental forms. The F_1 was like the elongate type. What is segregated (and hence genetically inherited) is the coefficient of the allometric growth keeping a precise and constant value throughout an entire growth period. From Sinnot, 1960.

Another interesting example of a morphogenetic constancy is a retention of a so-called conformal, or inverse symmetry, a notion introduced more than one hundred years ago and widely used in the modern physics (see, e.g., Kastrup, 1966) . This is a large and important case of symmetrical transformations which, by the way, includes all the abovementioned ones (that is, the rotational, reflectional and translational symmetry) as particular cases. From the morphogenetical point of view the most important property of the conformal symmetry is that it permits only those transformations of a body's form which are compatible with the constancy of the angles within a certain 'microgrid' inside a body. In the other words, the conformal symmetry preserves a small scale geometrical similarity permitting, at the same time, substantial deviations from such a similarity on a large scale level.

Some examples of the preservation or, better to say, restoration (after a certain delay) of the angular values on a microlevel have already been mentioned as related to what we have defined as a quasirelaxational strategy (2.3.6): what we mean is the cells' tendency to return towards isodiametrical shape after some temporary deviations from it. In accordance to the rules of a conformal symmetry, these restorations are closely linked with the changes of the higher levels geometry (as being the main causes of the latter ones). Meanwhile, the conformal symmetry is also perfectly maintained in a number of growth processes taking place in quite different species (Fig. 3.25). Its structural basis is most probably associated with some fundamental properties of a cytoskeleton and an extracellular matrix, namely with their tendencies to minimize cell surface areas and to deposit the fibrillar structures at right angles to each other. These properties can produce quite precise and robust morphologies even if the forces which cause macroscopic

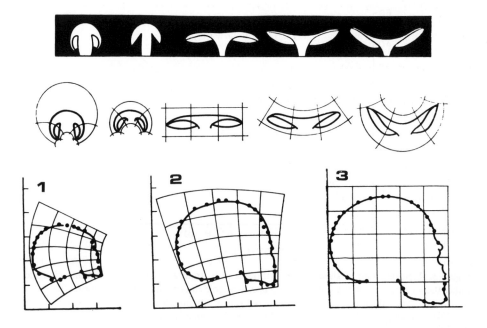

Figure 3.25. Preservation of a conformal symmetry in growth processes. Two upper rows: growth of a muschroom's fruit body. Lower row: growth of a human's skull. 1: a newborn infant, 2: 5 monthes old baby, 3: 20 years old individuum. In the both cases extensive changes of an overall morphology are associated with the preservation of 90° angular values (see grids). From Petuchov, 1981, with the author's permission.

deformations of a developing body are themselves arranged in a variable and not too precise manner. At the same time, their genetical basis well can be rather robust and non-specific. Hence, in the both above considered categories of morphogenetic processes, namely those characterized by the allometric growth and a maintenance of a conformal symmetry (which may be, by the way, quite well combined with each other) a parametrical role of genetical factors is well-defined and understandable, at least in general outlines. In most other cases however the developmental events assumed to be directly dependent upon genes are much more strictly localized, both in space and in time: such are, for example, almost all of the up to now described patterns of genes expression. Can such a precise locality be in any way combined with the parametrical (and thus largely smoothed) nature which we must ascribe to the primary genetical effects? This is a fundamental question which can be also reformulated as follows: *would it be possible to regard any finely localized genetical patterns as singularities of some continuous and invariable genetically parametrized holistic fields rather than a set of mutually independent properties?* We hope that the answer may be positive, otherwise our perspectives to get a satisfactory explanation of the local genetical effects will look rather gloomy. The roots of this our believing may be illustrated by the analysis of the model shapes shown in Fig. 2.34 (p. 143). Those shown, for example, in frames *b* and *d* can be indeed regarded as some kinds of 'point mutations' (or their phenocopies) in relation to their smoothed prototypes, *f* and *h* correspondingly, since *b* and *d* exhibit some local protuberances which are lacking in *f* and *h*. We know meanwhile that the both transformations (*f* → *b* as well as *h* → *d*) have been achieved by the changes of the mechanical parameters which are constant throughout the whole model rudiments and which are included into an invariable morphomechanical algorithm. Consequently, these 'point mutations' can be indeed regarded as the singularities of a holistic field smoothly modified by parametrical shifts. Another abovementioned case of a derivability of a local character from the global parametric changes was the existence and the attachement site of operculum in different hydroids species (p. 214): the both properties were shown to be dependent upon the overall patterns of growth pulsations. In order to know more about these extremely interesting global → local transformations a morphogenesis of the mutant samples, as compared with the wild type ones, should be studied with a much greater precision as it has been generally done. In few cases where such a description was enough detailized our expectations can be confirmed. Thus, in zebrafish mutants many abnormalities looking in 10 days embryos as local and mosaical ones appeared to be correlated with essentially holistic deviations of the dorsalwards convergent cell movements taking place at a gastrula stage (Müllins *et al.*, 1995).

Even in the best cases, meanwhile, such a reduction of any latest local characters to the earliest global features cannot be absolute. In the other words, as we have already mentioned beforehand, the genetically dependent parameters should definitely change their values in some, hopefully few, space–time points. How should we treat such a situation? Well, it is also not completely new for us: for several times we have already discussed the possibility of the parameters' drift with the rate even approaching that of the dynamic variables. The main question is now how this drift, or some more abrupt changes in the genetical parameters' values are regulated. The answers can be again either genocentrically or morphocentrically oriented. The first version implies that the genes can be regulated only by other genes, in their own (genomic) phase space, without employing any epigenetical factors unfolded within a real physical space–time of a developing embryo. In no way such an approach can be qualified as wrong or useless. On the contrary, some of its mostly refined and thoughtful modifications bring us, with the use of a sophisticated modelling, towards rather non-trivial conclusions about non-deterministic and even holistic properties of the genetic networks themselves (Kaufmann, 1974). However taken alone such an approach seems to be incomplete. If only recalling that the genes' expression patterns demonstrate a perfect 3-dimensional geometry (Shubin et al., 1997) adjusted according to the earlier established antero-posterior and dorso-ventral embryonic polarity one can hardly avoid the conclusion that the genetical parameters should depend somehow upon the epigenetic ones. The next acute question is whether these epigenetic influences are only short-range ones or they also include some long-range (holistic) factors. As argued beforehand, any patterns based exclusively upon the short-range interactions should be rather vulnerable and non-robust. Therefore, the assumption of the long-range epigenetic influences affecting the genetical parameters in some developmental space-time points at least, seems to be rather plausible.

By resuming this brief essay we may see quite real and promising perspectives for a cooperation between the 'morphocentric' and 'genocentric' approaches. My personal view is meanwhile that in order such a cooperation to become really fruitful the members of a recently dominating 'genocentric team' should make some important steps towards adopting and assimilating an entire system of notions associated with a morphocentric approach and borrowed, in the long run, from the self-organization theory. In particular, any remnants of a uniform determinism which showed already its inadequacy as applied to the hereditary–developmental relations should be definitely rejected and replaced by the idea of a parametric regulation. Until this being done, we are doomed to produce even more false problems and circulae viciosis without a hope to create a common and a rational theory of all these interlinked problems.

3.6. Summary of the chapter

Here we have tried to bring together those dynamic elements of the morphogenesis which we reviewed separately from each other in the previous chapter. In other words, we have made an attempt to interprete the real spatio-temporal order of the developmental events. We did so in the terms of a morphomechanics centered as previously around the idea of hyperrestoration of mechanical stresses.

We have explored several developmental successions. The first of them relates to the earliest stages of the egg development. We have analysed the events taking place prior to and during the first cleavage division cycle and have considered them as a kind of 'hyperrestoration swings' between the relaxation and retension of the ooplasmic membrane, with the involvement of some internal cytoskeleton mediated events. We have concluded that even this earliest developmental period is characterized by some regular and holistic morphomechanical events (in particular, the establishment of the stress singularities) which create the initial conditions for further morphogenesis proceeding from the blastula stage upwards.

Then we have given a morphomechanical reconstruction of the development of amphibian's embryos from blastula to the early tail–bud stage, with several discourses on the representatives of other related taxonomic groups. We have started from outlining the main structurally dynamical levels involved in morphogenesis and then came to the reconstruction proper, using again the idea of the stresses hyperrestoration as a unifying principle. We have also discussed the possible morphomechanical components of the genes expression and cell differentiation with a special emphasis to the phenomenae associated with embryonic inductions.

After this, we have made a great jump and came to lower invertebrates, namely to Hydrozoa belonging to Thecate suborder. Again, we have started from describing the main levels participating in their development. Then we have outlined a developmental archetype common for this group of animals and have interpreted the species-specific modulations of this archetype by regarding them as the functions of few developmental parameters which are opened, in principle, for genetical determination. In this way, a parametrical taxonomy for the Thecates have been suggested. Lastly, we have compared a 'morphocentric' approach employed throughout this book with a more ubiquitous 'genocentric' one and discussed the possibilities of a collaboration between these two trends.

CONCLUDING REMARKS

So, we have finished our travel throughout an 'unknown land' of organic development. What are our main impressions? The first of them would be, probably, that this land does not look to have been civilized very much: neither straight highways able to bring us within a known time period to a definite place, nor well equipped cross roads with detailed signposts do we see there. Nothing within this land goes on by strict schedule, nothing can be precisely predicted. True, after a close examination, you will notice in almost all the locations refined workshops, preparing precise products; but a subsequent transformation and a final distribution of these products is again a matter of considerable uncertainty. And nevertheless, after probably a brief embarrassment, you will feel an increased confidence and even a feeling of attraction to this strange land. Yes, you will find there neither any rigid instructions nor predetermined passages to this or that mountain top or river valley: but, in return, instead, a wide variety of different ways is opened for you, so that you are always free to select among them that one mostly suitable for you. A freedom, a robustness and a possibility of achieving, sooner or later, the desirable global goals are surprisingly coexistent with each other in this fairy kingdom. And probably it is more than a mere coincidence that the term 'field', meaning initially nothing more than a piece of land and only later being endowed with a deep physical interpretation, looks like the most suitable one for characterising the main properties of the 'morphogenetical land'. Instead of being a set of some rigid, highly specific points which can be enumerated and studied isolatedly from each other, or a network of deterministic connections linking these points in a one-to-one manner, it is a kind of a fluctuating spatio–temporal *continuum*, with no one of its localities being occupied by a certain indispensable and 'privileged' element. The heterogeneities which we see in this continuum comprise what are called *singularities*, that is some special (but exchangeable) entities, working as the attractors of the developmental trajectories. And in order to complete the picture, recall that this continuum contains several hierarchically coupled levels.

As has been already pointed out, this picture more or less correlates with some most deep and global modern concepts of the entire Universe, being emerged initially in several branches of physical sciences, such as a vacuum theory, quantum mechanics and a theory of self-organization (Bohm and Hiley, 1993; Prigogine, 1980; Laszlo, 1995). The main aim of this book was not so much to prolong this line of world wide generalizations as to demonstrate the desirability and even the necessity of this ideology for everyday research work in developmental biology. It seems quite obvious that the hitherto dominating opposite reductionist approaches, in spite of some impressive local achievements, involve the practice of research in ever more entangled *circulae viciosis*, without resolving at the same time the main questions, starting from this most naïve but inescapable one: "why so and so is happened in this very location and in this very time moment?"

What we really need today from the most general point of view is at least to complement, if not to exchange radically, a reductionist way of analysis inevitably directed towards a mosaic representation of a living system as a set of minor 'specific' units with another, implying the apprehension of an object as a really holistic and, at the same time, multilevel and dynamic entity. Now, instead of splitting a system into as small units as possible, our main task would be to understand its own internal 'language', as well as that of the associated 'contexts'. This is a difficult task which may be probably compared with that of a conductor dealing with a symphony's score: you should, at the same time, follow each single voice, try to appreciate its internal dynamics, not to miss its connections with other voices and, as a result, to perform the piece as an undivisible whole. Now, if we exchange 'voices' by 'levels', we just come to the formulation of the most general tasks of our multilevel approach. Such an approach would create a new mode of vision of living nature and, if properly used, will open to us, sooner or later, quite new horizons not only for theoretical science, but also for its reasonable applications.

Some first part of this way should be taken, in my view, along the road traced within the last three or so decades by the theory of self-organization: this is indeed the only well-elaborated physico-mathematical theory contiguous with morphogenesis, and at the present time it gives a sole possibility to transform the old romantic holistic desires (coming back to Goethe and naturphilosophers) into something belonging to a real science. Sooner or later, however, we shall feel ourselves somewhat burdened and limited by the physicalistic rigidity of this theory, as expressed, for example, in the sharp segregation of the different kinds of variables according to their characteristic times. As we know already, developing beings are much more fluent in this respect, being able not only to alter the T_{ch} -s of the parameters and the dynamic variables, up to their complete reversal, but even to change the entire set of variables. Along with that, the memory events and, in particular, the active Bergsonian memory, associated wuth different levels, plays in the prolonged chains of biological events a much greater role than in the purely physicalistic self-organizing models. Could all of this be brought together and incorporated into a kind of wider theory? Should we call it physical, or biological, or a hybrid theory? The future will tell.

After these rather general considerations, the set of notions which we have used throughout the book for describing morphogenetic events may look too narrow and primitive. Most of all, why did we pay so much attention to the mechanical stresses? Is it not another kind of mechanical dogmatism and an unpardonable simplification of the developmental phenomena? What we would emphasise first in this respect is that our interest in mechanical stresses was in no way invented artificially, nor came from some abstract considerations. Instead, in our research group it was originated more than two decades ago from some unexpected experimental results. At the present time, however, the outstanding role played by the mechanical stresses looks as though it is deeply rooted in the very essence of living beings. The reasons may be the follows.

Firstly, the long term persistence of the stressed states in developing embryos is closely linked with one of the mostly fundamental properties of the living matter, namely its stably maintained non-equilibricity, revealed by various chemokinetic, electrical, photonic, or other criteria. In order for this to become possible some part of the metabolic energy of the molecular and supramolecular structures should be stored (preserved from being dissipated) for long enough time periods. This is, however, just the same as if claiming that these structures are maintained during these time periods in the mechanically stressed states. Therefore, the existence of the mechanical stresses is not only a manifestation of a non-equilibricity, but also its main and necessary condition. Under these circumstances, one should not at all be surprised that the biological structures are mechanically stressed. What is far from being trivial and could not be expected *a priori*, is the effective non-dissipative transmission of the stressed states from the molecular–supramolecular level to the L_{ch} -s and T_{ch}-s several orders higher. Such a kind of transmission of some properties over several levels distance is typical for highly coherent systems, to which living ones belong according to many criteria (see Chapter 1, p. 53). On the other hand, that means that the stresses which we were dealing with might have a very complicated spatio–temporal structure: what we consider, in a rough approximation, as a constant homogeneous stress, may actually comprise an entire hierarchy of higher frequency oscillations, up to those elusive but attractive phenomena postulated by Fröelich (see, e.g., Fröelich and Kremer eds., 1983). Our present day approach to the mechanical stresses should be thus cautiously qualified as no more than a 'zero approximation', and a lot of important discoveries can be expected within the range of a 'cell's micromechanics', around and below the Frankel's barrier.

Returning now to the morphogenetic topics, we would like to emphasize the differences between the stress–shaping relations on the one hand and, for example, those which could take place between a shape and a chemical substance (see, in this respect, also Green, 1996a). In principle, as we know, any kind of a morphological structure can be prepatterned not only by a mechanical stresses, but also by a specific chemical substances. Meanwhile, for making possible such a chemical prepatterning, at least two kinds of events should take place. Firstly, you should establish in a very accurate and precise manner the proper parameters' values as well as the initial and border conditions for any subsequent morphogenetic event. Secondly – and this is a central point – for passing from a certain chemical substance to a shape dynamics, you should have an access to a highly specific 'coding–encoding machinery', establishing unambiguous relations between chemistry and geometry. On the other hand, both points become largely unnecessary insofar as you put aside the chemical prepatterns and address instead to mechanical stresses. The matter is that the latter already contain within themselves *ex definitio* a good piece of the shape dynamics in its overt form and do not require any special attributes for decoding it and extending this dynamics over the large distances. Or in other words, we may claim that by regarding morphogenesis as a stress–dependent process we bind together the physical space and the morphogenesis as close as possible, so that the latter can be regarded, in the long run, as the functions of some fundamental properties of a space. Another way to formulate almost the same idea is to claim that any morphogenetic process is to a greater extent 'natural', rather than artificially programmed, event. Accordingly, some notions widely used for describing the functioning of the man made programmed devices become largely inadequate as applied to the very essence of a morphogenesis. For example, one of the main strategies of a modern technology is to separate as far as possible the information blocks and the executive mechanisms; the opposite tendency should be considered as ineffective and even dangerous. Meanwhile, in the natural morphogenetic processes these two components look as being to a great extent merged with each other into an indivisible whole. It is of a fundamental importance to realize that the Nature tooks a largely 'non-technological', straight, robust and perhaps risky stress–mediated way of making shapes and structures instead of dealing with some strictly outlined pre-established programms and complicated coding-decoding mechanisms. In trying to borrow some examples of similar activities from human practice, one should recall sportif games, or the Paris-Dakkar rally (again an allegory of a wild land!) or, if seeking something more spiritual, an improvised creation of a masterpiece without a preliminary draft – but in no way a reproduction of a beforehand prepared text or an accurate unfolding of a pre-established programme. In any case, morphogenesis is a great creative game of a Nature, which is full of forces, ready to meet novelties and to make innovations not only during its evolution but also within each next individual life cycle.

This is not to say that much more deterministic tools are not used at all by the living matter: they are employed indeed, but for performing just the opposite tasks, as participating in the activities of some highly conservative and standardized 'product factories' preserved almost intact throughout the whole course of organic evolution. Such are the well known sets of homologous genes as well as the mechanisms of a signal transduction whose final results are meanwhile again very much context– and memory–dependent.

In such a way, by fluctuating back and forth between a morphogenetic creativity and standardized regulatory circuits, developing systems become able to link determinism to freedom, preciseness to variability, refinement to robustness and, in the long run, even to link human mind with the inorganic world. Recognizing these capabilities helps us to see the truth of Carl von Baer belief in the derivability of the "formative powers of the animal organisms" from the mostly fundamental laws of an overall world. And, considering the degree to which modern physics interprets its laws in terms of fields (rather than specific interaction of indiviual particles), why should biology continue to do the opposite?

REFERENCES

Alberts, B., Bray, D., Lewis, J. Raff, M., Roberts, K. and J.D.Watson (1989) *Molecular Biology of the Cell*, Garland Publishing Inc., N.Y., London.

Alt, W., and H. Kaiser (1994) Observations, modelling a simulation of keratinocyte movements, in N. Akkas (ed.) *Biomechanics of Active Movement and Division of Cells* (NATO ASI Series H: Cell Biology), Springer Verlag, Berlin, Heidelberg, pp. 445-452.

Anisov, A.M. (1991) *Time and Computer. A Non-geometrical Image of Time*, Nauka, Moskva (Russ).

Aon, M.A., and Cortassa, S.(1993) An allometric interpretation of the spatio-temporal organization of molecular an cellular processes, *Molec. Cell. Biochem.* **120**, 1-13.

Aon, M.A., and Cortassa, S. (1996) On the fractal nature of cytoplasm, *FEBS Letters* **344**, 1-4.

Aristotle (1975) *Metaphysics*, with an English translation by H. Tredernnick. London: William Heinemann; Cambridge (Mass.) Harvard Univ. Press, Vol. 1,2.

Arnolds, W.J.A., van den Biggelaar, J.A.M and Verdonk, N.H. (1983) Spatial aspects of cell interactions involved in the determination of dorsoventral polarity in equally cleaving Gastropods and regulative abilities of their embryos, as studied by micromere deleions in *Lymnaea* and *Patella*, *W. Roux' Arch. Dev. Biol.* **192,** 75-85.

Arshavsky, I.A. (1982) *Physiological mechanisms and regularities of the individual development*, Nauka, Moskva (Russ).

Baer, K.E. v. (1828) *Ueber Entwickelungsgeschichte der Thiere. Beobachtung und Reflexion. Erster Theil*. Konigsberg.

Banes, A.J., Tsuzaki, M., Yamamoto, J., Fischer, T., Brigman, B., Brown, T. and Miller, L. (1995) Mechanoreception at the cellular level: the detection, interpretation, and diversity of responses to mechanical signals, *Biochemistry and Cell Biology* **73**, 349-365.

Belintzev, B.N. (1991) *Physical Foundations of the Biological Morphogenesis*, Nauka, Moskva (Russ).

Belintzev, B.N., Beloussov, L.V. and Zaraisky, A.G. (1985) The model of epithelial morphogenesis basing on the elastic forces and contact polarization of cells. Biological consequences, *Ontogenez (Russ. J. Devel Biol.)* **16**, 437-449.

Belintzev, B.N., Beloussov, L.V. and Zaraisky, A.G. (1987) Model of pattern formation in epithelial morphogenesis, *J. theor. Biol.* **129**, 369-394.

Bellairs, R., Bromham, D.R. and Wylie, C.C. (1967) The influence of the area opaca on the development of the young chick embryo, *J.Embryol. exp. Morphol* **17**, 197-212.

Beloussov, L.V. (1973) Growth and morphogenesis of some marine Hydrozoa according to histological data and time-lapse studies, *Publ. Seto Mar. Biol. Lab.* **20**, 315-366.

Beloussov, L.V. (1978) Formation and cell structure of the tension lines in the axial rudiments of amphibians' embryos, *Ontogenez (Sov.J.Devel Biol.)* **9**, 121-131.

Beloussov, L.V. (1980) The role of tensile fields and contact cell polarization in the morphogenesis of amphibians axial rudiments, *Roux'Arch.Dev. Biol.* **188**, 1-7.

Beloussov, L.V. (1988) Contact polarization of the cells of *Xenopus laevis* embryos during gastrulation, 2. Morphogenetic and differentiational consequences of a relaxational cell polarization: relaxational morphoses, *Ontogenez (Sov.J.Devel Biol.)* **19**, 405-413.

Beloussov, L.V. (1994) The interplay of active forces and passive mechanical stresses in animal morphogenesis, in N. Akkas (ed.) *Biomechanics of Active Movement and Division of Cells* (NATO ASI Series H: Cell Biology), Springer Verlag, Berlin, Heidelberg, pp. 131-180.

Beloussov, L.V. and Bogdanovsky, S.B. (1980) Cell mechanisms of embryonic regulations in sea-urchin embryos, *Ontogenez (Sov.J.Devel Biol.)* **11**, 467-476.

Beloussov, L.V., Dorfman, J.G. and V.G. Cherdantzev (1975) Mechanical stresses and morphological patterns in amphibian embryos, *J. Embryol. Exp. Morphol.* **34**, 559-574.

Beloussov, L.V., Labas, J.A., Kazakova, N.I. (1993) Growth pulsations in hydroid polypes: kinematics, biological role and cytophysiology, in L. Rensing (ed.) *Oscillations and Morphogenesis*, Marcel Dekker, N.Y., Basel, Hong Kong, pp. 183-193.

Beloussov, L.V., Labas, J.A., Kazakova, N.I. and A.G. Zaraisky (1989) Cytophysiology of growth pulsations in hydroid polyps, *J. Exp. Zool.* **249**, 258-270.

Beloussov, L.V. and A.V. Lakirev (1988) Self-organization of biological morphogenesis: general approaches and topo-geometrical models, in I.Lamprecht, A.I.Zotin (eds) *Thermodynamics and Pattern Formation in Biology*, W. de Gruyter, Berlin, New York, pp. 321-336.

Beloussov, L.V. and A.V. Lakirev (1991) Generative rules for the morphogenesis of epithelial tubes, *J. Theor. Biol.* **152**, 455-468.

Beloussov, L.V., Lakirev, A.V. and I.I. Naumidi (1988) The role of external tensions in differentiation of *Xenopus laevis* embryonic tissues, *Cell Diff. & Devel.* **25**, 165-176.

Beloussov, L.V., Lakirev, A.V., Naumidi I.I. and V.V. Novoselov (1990) Effects of relaxation of mechanical tensions upon the early morphogenesis of *Xenopus laevis* embryos, *Int. J. Dev. Biol.* **34**, 409-419.

Beloussov, L.V. and Luchinskaia, N.N. (1983) A study of relay interactions of cells in the explants of amphibians' embryonic tissues, *Cytologia* **25**, 939-944 (in Russian).

Beloussov, L.V. and Luchinskaia, N.N. (1995) Biomechanical feedback in morphogenesis, as exemplified by stretch responses of amphibian embryonic tissues. *J. Biochem. Cell Biol.* **73**, 555-563.

Beloussov, L.V. and Luchinskaia, N.N. (1995a) Mechanodependent heterotopies of the axial rudiments in *Xenopus laevis* embryos. *Ontogenez (Russ.J.Devel Biol.)* **26**, 213-222.

Beloussov, L.V. and Naumidi, I.I. (1977) Contractility and epithelization in the axial mesoderm of chicken embryo, *Ontogenez (Sov.J.Devel Biol.)* **8**, 517-521.

Beloussov, L.V. and Naumidi, I.I. (1983) Cell contacts and rearrangements preceeding somitogenesis in chick embryo, *Cell Differentiation* **12**, 191-204.

Beloussov, L.V. and K.V.Petrov (1983) The role of cell interactions in the differentiation of the induced tissues of amphibian embryos, *Ontogenez (Sov.J.Devel Biol.)* **14**, 21-29.

Beloussov, L.V., Saveliev, S.V., Naumidi, I.I. and V.V.Novoselov (1994) Mechanical stresses in embryonic tissues: patterns, morphogenetic role and involvement in regulatory feedback, *Int. Rev. Cytol.* **150**, 1-34.

Beloussov, L.V. and Snetkova, E.V. (1994) The dependence of the differentiation potencies of the marginal zone areas in the early gastrulae of *Xenopus laevis* upon their morphogenetic movements. *Ontogenez (Russ.J.Devel Biol.)* **25**, 63-71.

Beloussov, L.V. and Zhadan, A.L. (1993) Morphological reactions of Echinodermata embryos upon the decrease of an osmotical gradient between blastocoel and external environment, *Ontogenez (Russ.J.Devel Biol.)* **24**, 32-36.

Bereiter-Hahn, J. (1987) Mechanical principles of architecture of eukaryotic cells, in J. Bereiter-Hahn, O.R.Anderson and W.-E. Reif eds, *Cytomechanics*, Springer Verlag, Berlin, Heidelberg, pp. 3-30.

Bereiter-Hahn, J. and H. Lüers (1994) The role of elasticity in the motile behaviour of cells, in N. Akkas (ed.) *Biomechanics of Active Movement and Division of Cells* (NATO ASI Series H: Cell Biology), Springer Verlag, Berlin, Heidelberg, pp. 181-230.

Bergmann, J.E., Kupfer, A. and Singer, S.J. (1983) Membrane insertion at the leading edge of motile fibroblasts, *Proc. Nat. Acad. Sci USA* **80**, 1367-1371.

Bergson, H. (1896) *Matiere et Memoire. Essai sur relation du corps a l'exprit*, Ancienne librairie Germen Bailliere et C-ie, Paris.

Betchaku, T. and J.P.Trinkaus (1978) Contact relations, surface activity, and cortical microfilaments of marginal cells of the enveloping layer and of the yolk syncytial and yolk cytoplasmic layers of *Fundulus* before and during epiboly, *J. Exp. Zool.* **206**, 381-426.

Betchaku, T. and J.P.Trinkaus (1986) Programmed endocytosis during epiboly of *Fundulus heteroclitus*, *Amer. Zool.* **26**, 193-199.

Bistolfi, F. (1991) *Biostructures and Radiation Order-Disorder*, Edizioni Minerva Medica, Torino 1991.

Bleschmidt, E., Gasser, R.F. (1978) *Biokinetics and Biodynamics of Human Differentiation*, Charles C. Thomas Publisher, Springfield Illinois USA.

Bluemenfeld, L.A. (1983) *Physics of Bioenergetic Processes*, Springer Verlag, Berlin.

Bode, P.M. and Bode, H.R. (1984) Formation of pattern in regenerating tissue pieces of *Hydra attenuata*, III. The shaping of the body column, *Devel Biol.* **106**, 315-325.

Bohm, D. (1980) *Wholeness and the Implicate Order*, Routledge and Kegan Paul, London.

Bohm, D. and B.J.Hiley (1993) *The undivided universe*, Routledge and Kegan Paul, London.

Braam, J. and Davis, R.W. (1990) Rain-, wind- and touch-induced expression of calmodulin and calmodulin-related genes in *Arabidopsis, Cell* **60**, 357-364.

Brunette, D.M. (1984) Mechanical stretching increases the number of epithelial cells synthesizing DNA in culture, *J. Cell Sci.* **69**, 35-45.

Carroll, S.B. (1995) Homeotic genes and the evolution of arthropods and chordates, *Nature* **376**, 479-485.

Cherdantsev, V.G. (1977) Spatial unfoldings of the morphogenetic movements as the elements of the oral field in amphibian embryos, *Ontogenez (Sov.J.Devel Biol.)* **8**, 335-360.

Cherdantsev, V.G. and V.A.Skobeyeva (1994) The morphological basis of self-organization. Developmental and evolutionary aspects, *Rivista di Biologia (Biology Forum)* **87**, 57-85.

Cherdantseva, E.V. and Cherdantsev, V.G. (1985) Determination of a dorso-ventral polarity in the embryos of a fish Brachydanio rerio (Teleostei), *Ontogenez (Sov.J.Devel Biol.)* **16**, 270-280.

Chrzanowska-Wodnicka, M., and K. Burridge (1996) Rho-stimulated contractility drives the formation of stress fibers and focal adhesions, *J.Cell Biol.* **133**, 1403-1415.

Cohen, M.S. (1977) The cyclic AMP control system in the development of *Dictyostelium discoideum*, *J. Theor. Biol.* **69**, 57-86.

Condic, M.L., D. Fristrom and J.W.Fristrom (1991) Apical cell changes during Drosophila imaginal leg disc elongation: a novel morphogenetic mechanism, *Development* **111**, 23-33.

Coulombre, A.J. (1956) The role of intraocular pressure in the development of the chick eye, *J. exp. Zool.* **133**, 211-226.

Crick, F.C.H. and P.A.Lawrence (1975) Compartments and polyclones in insect development, *Science* **189**, 340-347.

Cummings, F.W. (1994) Aspects of growth and form, *Physica D* **79**, 146-163.

Curie, P. (1894) De symmetrie dans les phenomenes physiques: symmetrie des champs electriques et magnetiques, *J. de Physique* **Ser. 3**, 393-427.

Dale, L. and J.M.W.Slack (1987) Fate map for the 32-cell stage of *Xenopus laevis, Development* **99**, 527-551.

Danilchik, M.V. and J.M.Denegre (1991) Deep cytoplasmic rearrangements during early development in *Xenopus laevis, Development* **111**, 845-856.

Dartsch, P.C. and Hammerle, H. (1986) Orientation response of arterial smooth muscle cells to mechanical stimulation, *Europ. J. Cell Biol.* **41**, 339-346.

Dawid, I.B. (1994) Intercellular signalling and gene regulation during early embryogenesis of *Xenopus laevis, J. Biol. Chem.* **269**, 6259-6262.

Dennerly, T.J., Lamoureux, P., Buxbaum, R.E. and Heidemann, S.R. (1989) The cytomechanics of axonal elongation and retraction, *J. Cell Biol.* **109**, 3073-3083.

Detlaff, T.A. (1977) Establishment of the organization of a germinated egg in amphibians and fishes, in *Modern Problems of Oogenesis*, T.A. Detlaff ed. Nauka, Moskva pp. 99-144 (in Russian).

Devreotes, P.S., Potel, M.J. and Machay, S.A. (1983) Quantitative analysis of cAMP waves mediating aggregation in Dictyostelium discoideum, *Devel Biol.* **96**, 405-415.

Dohmen, M.R. and J.C.A. van der Mey (1977) Local surface differentiations at the vegetal pole of the eggs of *Nassarius reticulatus, Buccinum indatum*, and *Crepidula fornicata* (Gastropoda, Prosobranchia), *Devel Biol.* **61**, 104-113.

Doniach, T. (1992) Induction of anteroposterior neural pattern in *Xenopus* by planar signals, *Development (Suppl)*, 183-193.

Dorfman, Ja.G. and V.G.Cherdantzev (1977) Structure of morphogenetic movements of gastrulation in Anura, I. Destabilization of ooplasmic segregation and cleavage by clinostating, *Ontogenez (Sov.J.Devel Biol.)* **8**, 238-250.

Driesch, H. (1892) Entwicklungsmechanische Studien. I. Der Wert der beiden ersten Furchungszellen in der Echinodermenentwicklung. Experimentelle Erzeugung von Theil- und Doppelbildungen, *Z. Wiss. Zool.* **53**, 160-178.

Driesch, H. (1921) *Philosophie des Organischen*, Engelmann, Leipzig,

Durston, A.J. (1973) *Dictyostelium discoideum* field as excitable media, *J. Theor. Biol.* **42**, 483-504.

Elsdale, T. (1972) Pattern formation in fibroblast cultures: an inherently precise morphogenetic process, in C.H.Waddington (ed.) *Towards a Theoretical Biology* 4. *Essays,* Edinburgh University Press, Edinburgh, pp. 95-108.

Ettensohn, C. A. (1985) Gastrulation in the sea urchin embryo is accompanied by the rearrangement of invaginating epithelial cells, *Devel. Biol.* **112**, 383-390.

Evans, E.A. and Skalak, R. (1980) *Mechanics and Thermodynamics of Biomembranes*, CRC Press, Inc. Boca Raton, Florida.

Fink, R.D. and M.S.Cooper (1996) Apical membrane turnover is accelerated near cell-cell contacts in an embryonic epithelium, *Devel Biol.* **174**, 180-189.

Fleischner, M., Wohlfarth-Bottermann, K.E. (1975) Correlations between tension force generation, fibrillogenesis and ultrastructure of cytoplasmic actomyosin during isometric standarts, *Cytobiologie*, **10**, 339-365.

Forgacs, G. (1995) On the possible role of cytoskeletal filamentous networks in intracellular signaling: an approach based on percolation, *J. Cell Sci.* **108**, 2131-2143.

Frankel, J. (1989) *Pattern Formation. Ciliate Studies and Models,* Oxford University Press, N.Y., Oxford.

Fristrom, D. (1976) The mechanism of evagination of imaginal discs of *Drosophila melanogaster* III. Evidence for cell rearrangement, *Devel Biol.* **54**, 163-171.

Fröhlich, H. and Kremer F., eds (1983) *Coherent Excitations in Biological Systems*, Springer -Verlag, Berlin.

Galau, G.A., Klein, W.H., Davis, M.M., Wold, B.J., Britten, R.J. and Davidson, E.H. (1976) Structural gene sets active in embryos and adult tissues of the sea urchin, *Cell* 7, 485-505.

Georgiev, G.P., and Beloussov, L.V. (1986) Role of a short-termed disturbance of cell contacts upon the differentiation of a mesoderm of *Xenopus laevis* embryos, *Ontogenez (Sov. J. Devel. Biol.)* **17**, 84-86.

Gerhart, J., J.-P.Vincent, S.Scharf and B.Rowning (1988) Cortical-subcortical rotation in the amphibian egg, in *Signal Transduction in Cytoplasmic Organization and Cell Motility*, Alan R.Liss Inc, pp. 245-250.

Gerisch, G., Maeda, Y., Malshov, D., Roos, W., Wick, U.,Wurster, B (1977) Cyclic AMP signals and the control of cell aggregation in *D. discoideum*, in *Development and Differentiation in Cellular Slime Moulds*, Elsevier, North Holland Biomedical Press, pp. 105-124.

Gilbert, S.F., Opitz, J.M. and Raff, R.A. (1996) Resynthesizing evolutionary and developmental biology, *Devel. Biol.* **173**, 357-372.

Godsave, S.F. and J.M.W.Slack (1991) Single cell analysis of mesoderm formation in the Xenopus embryo, *Development* **111**, 523-530.

Goodwin, B.C. (1985) What are the causes of morphogenesis? *BioEssays* **3**, 32-36.

Goodwin, B.C. (1994) *How the Leopard Changed Its Spots. The Evolution of Complexity,* Weidenfeld and Nicolson, London.

Goodwin, B.C. and Briere, C. (1994) Mechanics of the cytoskeleton and morphogenesis of *Acetabularia*, *Int. Rev. Cytol.* **150**, 225-242.

Goodwin, B.C. and Trainor, L.E.H. (1985) Tip and whorl morphogenesis in *Acetabularia* by calcium-regulated strain fields, *J. Theor. Biol.* **117**, 79-106.

Graff, J.M.(1997) Embryonic patterning: to BMP or not to BMP, that is the question, *Cell* **89**, 171-174.

Green, J.B.A., New, H.V., and Smith, J.C. (1992) Responses of embryonic *Xenopus* cells to activin and FGF are separated by multiple dose thresholds and correspond to distinct axes of the mesoderm, *Cell* **71**, 731-739.

Green, P. (1996) Transductions to generate plant form and pattern: an essay on cause and effect, *Annals of Botany* **78**, 269-281.

Green, P. (1996a) Expression of form and pattern in plants - a role for biophysical fields, *Cell and Develop. Biol.* **7**, 903-911.

Green P., Steele, C.S. and S.C.Rennich (1996) Phyllotactic patterns: a biophysical mechanism for their origin. *Annals of Botany* **77**, 515-527.

Gurdon, J (1988) A community effect in animal development, *Nature* **336**, 772-774.

Gurdon, J.B., Harger, P., Mitchell, A. and Lemaire, P. (1994) Activin signalling and response to a morphogen gradient, *Nature* **371**, 487-492.

Gurwitsch, A.G. (1914) Der Vererbungsmechanismus der Form, *Arch. Entwicklungsmech. des Organismen* **39**, 516-577.

Gurwitsch, A.G. (1922) Über den Begriff des embryonalen Feldes, *Arch. Entwicklungsmech. des Organismen* **51**, 383-415.

Gurwitsch, A. G. (1930) *Die histologischen Grundlagen der Biologie*, Fisher, Jena.

Gurwitsch, A.G. (1944) *A theory of the Biological Field*, Moskva, Sovetskaya Nauka (in Russian).

Gustafson, T. and L. Wolpert (1967) Cellular movements and contacts in sea urchin morphogenesis, *Biol.Rev.* **42**, 442-498.

Haken, H. (1988) Morphogenesis of behaviour and information compression in biological systems, in I. Lamprecht and A.I. Zotin (eds.), *Thermodynamics and Pattern Formation in Biology*, W. de Gruyter, Berlin, N.Y., pp. 321-336.

Haraway, D. (1976) *Crystals, Fabrics and Fields: Metachors of Organicism in Twentieth Century Developmental Biology*, Yale University Press, New Haven.

Hardin, J.D. and Cheng, L.Y. (1986) The mechanisms and mechanics of archenteron elongation during sea urchin gastrulation, *Devel. Biol.* **115**, 540-501.

Harold, F.M. (1990) To shape a cell: an inquiry into the causes of morphogenesis of microorganisms, *Microbiol. Rev.* **545**, 381-349.

Harris, A.K. (1990) Testing cleavage mechanisms by comparing computer simulations to actual experimental results, *Ann N.Y. Acad. Sci* **582**, 60-77.

Harris, A.K. (1994) Locomotion of tissue culture cells considered in relation to ameboid locomotion, *Int. Rev. Cytol.* **150**, 35-68.

Harris, A.K. (1994a) Cytokinesis: the mechanism of formation of the contractile ring in animal cell division, in N. Akkas (ed.) *Biomechanics of Active Movement and Division of Cells* (NATO ASI Series H: Cell Biology), Springer Verlag, Berlin, Heidelberg, pp. 37-66.

Harris, A.K., Stopak, D. and Warner, P. (1984) Generation of spatially periodic patterns by a mechanical instability: a mechanical alternative to the Turing model, *J. Embryol. Exp. Morphol.* **80**, 1-20.

Harris, A.K., Wild, P. and Stopak, D. (1980) Silicone rubber substrata: a new wrinkle in the study of cell locomotion, *Science* **208**, 177-179.

Hemmati-Brivanlou, A., Melton, D. (1997) Vertebrate embryonic cells will become nerve cells unless told otherwise, *Cell* **88**, 13-17.

Ho, M.W. (1993) *The Rainbow and the Worm. The Physics of Organisms*, World Scientific, Singapore.

Holtfreter, J. (1933) Der Einfluss der Wirtsalter und vershiedenen Organbezirken auf Differenzierung von angelagerten Gastrulaektoderm, *Arch. Entw.-Mech. Org.* **127**, 619-775.

Holtfreter, J. (1938a) Differenzierungspotenzen isolierter Teile der Urodelengastrula, *Arch. Entw.-Mech. Org.* **138**, 522-656.

Holtfreter, J. (1938b) Differenzierungspotenzen isolierter Teile der Anurengastrula, *Arch. Entw.-Mech. Org.* **138**, 657-738.

Horstadius, S. (1939) The mechanics of sea-urchin development studied by operative methods, *Biol. Rev* **14**, 132-179.

Houliston, E. and R.P.Elinson (1991) Evidence for the involvement of microtubules, ER, and kinesin in the cortical rotation of fertilized frog eggs, *J. Cell Biol.* **114**, 1017-1028.

Huxley, J.S. (1932) *Problems of Relative Growth*, Methuen, London.

Ilizarov, G.A. (1984) Stretching stress as a factor exciting and supporting the regeneration and growth of the bone's and soft tissues, in *Structure and Biomechanics of the Skeletal-Muscular and Heart-Vessels Systems of Vertebrates*, Kiev, Ukraine Rep. Conference, pp.38-40.

Ingber, D.E., Karp, S., Plopper, G., Hansen, L. and Mooney, D. (1993) Mechanochemical transduction across extracellular matrix and through the cytoskeleton. In *Physical Forces and the Mammalian Cell* (ed. J.A. Frangos) Acad. Press N.Y. pp. 61-79.

Ingber, D.E., Dike, L., Hansen, L., Karp, S., Liley, H., Maniotis, A., McNamee, H., Mooney, D., Plopper, G., Sims, J., Wang, N. (1994) Cellular tensegrity: exploring how mechanical changes in the cytoskeleton regulate cell growth, migration and tissue pattern during morphogenesis, *Int. Rev. Cytol.* **150**, 173-224.

Isaeva, V.V. (1984) On the morphogenetical role of a cortical cytoskeleton and plasmatic membrane of an egg cell, *Cytologia* **26**, 5-13 (Russ).

Isaeva, V.V., Presnov, E.V. (1990) *Topological Structure of Morphogenetic Fields*, Moskva, Nauka (Russ).

Ivanenkov, V.V., Minin, A.A., Meschneryakov, V.N. and Martynova, L.E. (1987) The effect of local cortical microilament disorganization on ooplasmic segregation in the loach (*Misgurnus fossilus*) egg, *Cell Differentiation* **22**, 19-28.

Ivanenkov, V.V., Meschneryakov, V.N. and Martynova, L.E. (1990) Surface polarization in loach eggs and two-cell embryos: correlation between surface relief, endocytosis and cortex contractility, *Int. J. Dev. Biol.* **34**, 337-349.

Ivanova-Kazas, O.M. (1978) *Comparative Embryology of Invertebrates. Echinodermata and Hemichordata.*, Nauka, Moskva (Russ).

Jaffe, L.F. (1968) Localization in the developing *Fucus* eggs and the general role of localizing currents, *Advances in Morphogenesis* **7**, 295-328.

Jaffe, L.F. (1969) On the centripetal course of development, the *Fucus* egg, and self-electrophoresis, *Devel. Biol. Suppl.* **3**, 83-111.

Jaffe, L.F. (1981) The role of ionic currents in establishing developmental pattern, *Phil. Trans. R. Soc. London B* **295**, 553-566.

Jaffe, L.F. (1993) Classes and mechanisms of calcium waves, *Cell Calcium* **14**, 738-745.

Jeffery, W.R. (1992) A gastrulation center in the ascidian egg, *Development* Suppl *(C.D. Stern and P.W.Ingham eds)* pp. 53-63.

Johnston, M.H. (1981) Membrane events associated with the generation of a blastocyst, *Int. Rev. Cytol., suppl.* **12**, 1-37.

Jones, D., Leivseth, G. and J. Tenbosch (1995) Mechano-reception in osteoblast-like cells, *Biochemistry and Cell Biology* **73**, 525-534.

Kaneko, K. and T. Yomo (1997) Isologous diversification: a theory of cell differentiation, *Bull. Math. Biol.* **59**, 139-196.

Karlsson, J. (1984). Morphogenesis and compartments in Drosophila. In: *Pattern Formation. A Primer in Developmental Biology* (G.M.Malacinski ed.) McMillan Publ Co. N.Y., L. pp. 323-337.

Kastrup, H.A. (1966) Conformal group in space-time, *Physic. Rev.* **142**, 1060-1071.

Kauffman, S.A. (1974) The large-scale structure and dynamics of gene control circuits: an ensemble approach, *J. Theor. Biol.* **44**, 167-190.

Kazakova, N.I. Kossevitch, I.A., Beloussov, L.V. (1997). Effects of the mechanical deformations and the cytoskeletal inhibitors upon growth pulsations in hydroid polypes. *Ontogenez (Russ. J. Devel Biol)* **28**,

Kazakova, N.I., Labas, Ju.A. and Beloussov, L.V. (1991). A correlaion between the polarity of electrical reactions and the distoproximal polarity in hydroid polypes. *Ontogenez (Sov.J.Devel Biol.)* **22**, 84-89.

Kazakova, N.I., Zierold, K., Plickert, G., Labas, Ju.A. and Beloussov, L.V. (1994). X-ray microanalysis of ion contents in vacuoles and cytoplasm of the growing tips of a hydroid polyp as related to osmotic changes and growth pulsations. *Tissue & Cell* **26**, 687-697.

Kazakova, N.I., Jones, D.B., Zierold, K., Plickert G., Beloussov, L.V (1995) Ionic dynamics and membrane potential oscillations during growth pulsations in hydroid polypes. *Abstracts of 6th International Workshop on Hydroid Development Evangelische Akademie Tutzing* p. 80.

Keller, R.E. (1978) Time-lapse cinemicrographic analusis of superficial cell behavior during and prior to gastrulation in *Xenopus laevis, J. Morph.* **157**, 223-248.

Keller, R.E. (1987) Cell rearrangements in morphogenesis, *Zool. Sci* **4**, 763-779.

Keller, R.E. and J. Hardin (1987) Cell behaviour during active cell rearrangement: evidence and speculations, *J.Cell Sci Suppl* **8**, 369-393.

Keller, R.E. and P. Tibbetts (1989) Mediolateral cell intercalation in the dorsal axial mesoderm of *Xenopus laevis, Devel Biol* **131**, 539-549.

Keller, R. and J. Trinkaus (1987) Rearrangement of enveloping layer cells without disruption of the epithelial permeability barrier as a factor in *Fundulus* epiboly, *Dev. Biol.* **120**, 12-24.

Kochav, S. and Eyal-Giladi, H. (1971) Bilateral symmetry in chick embryo determination by gravity, *Science* **171**, 1027-1029.

Kolega, J. (1986) Effects of mechanical tension on protrusive activity and microfilament and intermediate filament organization in an epidermal epithelium moving in culture, *J. Cell Biol.* **102**, 1400-1411.

Komuro Issei, et al. (1991). Mechanical loading stimulates cell hypertrophy and specific gene expression in cultured rat cardiac myocytes: possible role of protein kinase C activation. *J.Biol. Chem.* **266**, 1265-1268.

Kortmulder, K. (1994) Towards a field theory of behaviour, *Acta Biotheoretica* **42**, 281-293.

Korver, G.H.V., van de Stadt, R.J., van Kampen, G.P.J and van der Korst, J.K. (1990). Cyclic compression of cultured anatomically intact articular cartilage stimulates synthesis of small proteoglycan. *Eur.J.Cell Biol.* **53**, Suppl N 31, 21.

Kossevitch, I.A. (1990) Development of the peduncules' and stolons' internodes in hydroids of *Obelia* genus, *Vestnik Mosk. Universiteta* Ser. **16** (Biologia) **3**, 26-32 (in Russian).

Kossevitch, I.A. (1996) Regulation of formation of the elements in the hydroid polypes colony, *Ontogenez (Russ. J. Devel Biol)* **27**, 114-121.

Krauss, Ju.A., Cherdantzev, V.G. (1995) A primary epithelization of cells in the early ontogenesis of a marine hydroid, *Dynamena pumila L. Ontogenez (Russ J.Devel.Biol.)* **26**, 223-230.

Krinsky, V.I., Zhabotisky, A.M. (1981). Autowave structures and the perspectives of their investigations, in M.T. Grechova (ed.) *Autowave Processes in Diffusional Systems* Gorky, Inst. Applied Physics Acad. Sci. USSR, pp. 6-32 (Russ).

Kropf, D.L. (1994) Cytoskeletal control of cell polarity in a plant zygote, *Devel. Biol.* **165**, 361-371.

Kucera, P., Monnet-Tschudi, F. (1987). Early functional differentiation in the chick embryonic disc: interactions between mechanical activity and extracellular matrix. *J. Cell Sci Suppl.* **8**, 415-432.

Labas, Ju.A., Beloussov, L.V., Kazakova, N.I. and Badenko, L.A. (1987). The reactions to electric fields as an indication of the relation between the organismic an the cell polarity in hydroid polypes. *Ontogenez (Sov.J.Devel Biol.)* **18**, 154-168.

Langer, J.S. (1989). Dendrites, viscous fingers, and the theory of pattern formation. *Science* **243**, 1150-1156.

Larsson T., Aspden, R.M., Heinegard, D. (1991). Effects of mechanical load on cartilage matrix biosynthesis in vitro. *Matrix* **11**, 388-394.

Laszlo, E. (1995). *The interconnected universe.* Conceptual foundations of transdisciplinary unified theory. World Scientific, Singapore, New Jersey, London, Hong Kong.

Lawrence, P.A. (1992). *The Making of a Fly. The Genetics of Animal Design.* Blackwell Scientific Publications L., etc.

Lecuit, T. and S.M. Cohen (1997). Proximal-distal axis formation in the *Drosophila* leg. *Nature* **388**, 139-145.

Levin, S.V., Korenstein, R. (1991). Membrane fluctuations in erythrocytes are linked to MgATP dependent dynamic assembly of the membrane skeleton, *Biophysic. J.* **60**, 733-737.

Levina, E.M., Domnina, L.V., Rovensky, Y.A., Vasiliev, J.M. (1996). Cylindrical substratum induces different patterns of actin microfilament bundles in nontransformedand in ras-transformed epitheliocytes. *Exptl Cell Res* **229**, 159-165.

Lockhart, J.A. (1965). An analysis of irreversible plant cell elongation, *J. Theor. Biol.* **8**, 264-275.

Maniotis, A.J., Chen, C.S., Ingber, D.E. (1997). Demonstration of mechanical connections between integrins, cytoskeletal filaments, and nucleoplasm that stabilize nuclear structure. *Proc. Natl Acad. Sci. USA* **94**, 849-854.

McClay, D.R. and C.Y. Logan (1996). Regulative capacity of the archenteron during gastrulation in the sea urchin, *Development* **122**, 607-616.

Martynov, L.A. (1982). The role of macroscopic processes in morphogenesis, in A.I.Zotin, E.V.Presnov (eds), *Mathematical Biology of Development*, Nauka, Moskva, pp. 135-154 (Russ).

Meinhardt, H. (1982) *Models of Biological Pattern Formation*, Academic Press, N.Y., L.

Meshcheryakov, V.N. (1983) Imitation of a spiral cytotomy by membrane ghosts of the pond snails zygotes, *Ontogenez (Sov. J. Devel. Biol.)* **14**, 82-85.

Mescheryakov, V.N. (1991) How genes distinguish right and left? in E.V.Presnov, V.M.Maresin, A.I.Ivanov (eds), *Analytical Aspects of Differentiation*, Nauka, Moskva, pp. 137-166 (Russ).

Michailov, A.S., Loskutov, A.Ju. (1990) *An Introduction to the Synergetics*. Moskva, Nauka 270 p. (in Russian).

Middleton, C.A. (1977) The effects of cell-cell contact on the spreading of pigmented retina epithelial cells in culture, *Exp. Cell Res.* **109**, 349-359.

Mochitate, K., Pawelek, P., Grinnell, F. (1991) Stress relaxation of contracted collagen gels: disruption of actin filament bundles, release of cell surface fibronectin and down-regulation of DNA and protein synthesis, *Expt Cell Res* **193**, 261-265.

Moore, A.R. (1941) On the mechanisms of gastrulation in Dendraster excentricus, *J. Exp. Zool.* **87**, 101-111.

Müllins *et al* (1995) Genes establishing dorsoventral pattern formation in the zebrafish embryo: the ventral specifuing genes. *Development* **123**, 81-93.

Nasmyth, K. and Jansen, R.-P. (1997) The cytoskeleton in nRNA localization and cell differentiaion, *Curr. Op. Cell Biol.* **9**, 396-400.

Nédélec, F.J., Surrey, T., Maggs, A.C., Leibler, S. (1997) Self-organization of microtubules and motors, *Nature* **389**, 305-308.

Nicolis, G., and Prigogine, I. (1977) *Self-organization in non-equilibrium systems*. Wiley, N.Y.

Nieuwkoop, P.D. (1977). Origin and establishment of an embryonic polar axis in amphibian development. *Curr. Top. Devel. Biol.* **11**, 115-117.

Neyfach, A.A., Timofeyeva, M.Ja. (1978). *Problems of regulation in the molecular biology of development*. Moskva, Nauka (in Russian).

Nijhout, H.F. (1990) Metaphors and the role of genes in development, *BioEssays* **12**, 441-446.

Nüsslein-Volhard, Chr., H.R. Frohnhöfer, R. Lehman (1987). Determination of anteroposterior polarity in *Drosophila. Science* **238**, 1675-1681.

Odell, G.M. (1984). A mathematically modelled cytogel cortex exhibits periodic Ca^{++}-modulated contraction cycles seen in *Physarum* shuttle streaming. *J. Embryol. Exp. Morphol.* **83** (Suppl), 261-287.

Odell, G.M., Oster, G., Alberch, P., Burnside, B. (1981). The mechanical basis of morphogenesis. I. Epithelial folding and invagination, *Devel Biol* **85**, 446-462.

Opas, M. (1994). Substratum mechanics and cell differentiation. *Int. Rev. Cytol.* **150**, 119-138.

Opitz, J.M. (1993) Blastogenesis and the "primary field" in human development, *Birth Defects: Original Article Series* **29**, 3-37.

Oster, G.F. (1984) On the crawling of cells. *J. Embryol. Exp. Morphol.* **83** (Suppl), 329-364.

Oster, G.F., Murray, J.D., Harris, A.K. (1983) Mechanical aspects of mesenchymal morphogenesis. *J. Embryol. Exp. Morphol.* **78**, 83-125.

Oster, G.F., Murray, J.D., Odell G.M. (1985) The formation of microvilli, in *Molecular Determinants of Animal Form (UCLA Symposium on Molec. And Cell Biol., New Series*, **31**) Alan R. Liss Inc, pp. 365-384.

Oster, G.F. and G.M.Odell (1984). Mechanics of cytogels, I: Oscillations in Physarum. *Cell Motility* **4**, 469-503.

Pattee, H.H. (ed) (1973) *Hierarchy Theory. The Challenge of Complex Systems* (The Internat. Library of Systems Theory and Phylosophy, XV). Brazillers, N.Y. 156 pp.

Peskin, Ch.S., G.M. Odell and G.F. Oster (1993) Cellular motions and thermal fluctuations: the Brownian ratchet, *Biophys. J.* **65**, 316-324.

Petrov, K.V. and Beloussov, L.V. (1984) A kinetic of a contact cell polarization in the induced tissues of amphibian embryos, *Ontogenez (Sov.J.Devel Biol.)* **15**, 643-648.

Petuchov, S.V. (1988) *Geometries of a living nature and the algorithms of a self-organization*, Znanie, Moskva (in Russian).

Plickert, G. (1980) Mechanically induced stolon branching in *Eirene viridula* (Thecata, Campanulinidae), in P. Tardent and R. Tardent eds, *Developmental and Cellular Biology of Coelenterates*, Elsevier/North Holland, pp. 185-193.

Pokrywka, N.J., Stephenson, E.C. (1991) Microtubules mediate the localization of bicoid mRNA during Drosophila oogenesis. *Development* **113**, 55-66.

Polezhayev, A.A., Ptytsin, M.O.(1991) Formation of spatial structures and the differentiation in the bacterial colonies, in E.V.Presnov, V.M.Maresin, A.I.Ivanov (eds.), *Analytical Aspects of Differentiation* Nauka, Moskva, pp. 167-180 (Russ).

Popp, F.-A., K.H. Li and Q. Gu (eds) (1992) *Recent Advances in Biophoton Research and its Applications*, World Scientific, Singapore.

Preston, R.D. (1974), *The Physical Biology of Plant Cell Walls*, Chapman and Hall, London.

Prigogine, I. (1980) *From Being to Becoming. Time and Complexity in the Physical Sciences.* W.H. Freeman and Co, N.Y.

Raff, R.A. and Kaufman, T.C. (1983) *Embryos, Genes and Evolution*. Macmillan Publ. Co, N.Y., L.

Resnick, N., Collins, T., Atkinson, W., Bonthron, D.T., Dewey, C.F., Jr, and Gimbrone, M.A., Jr (1993), Platelet-derived growth factor B chain promoter contains a cis-acting fluid shear-stress-responsive element, *Proc. Natl Acad. Sci. USA* **90**, 4591-4595.

Robertson, A., Grutsch, J.F. (1981) Aggregation in Dictyostelium discoideum. *Cell* **24**, 603-611.

Romanovsky, Ju. M., Stepanova, N.V., Chernavsky, D.S. (1984), *Mathematical Biophysics*, Nauka, Moskva (in Russian).

Rosales, O., Shin, T., Sumpid, B.E. (1990). Acute change in cyclic stretch frequency in vitro stimulates production of diacylglycerol and inositol phosphate. *J. Cell Biol.* **111**, 84.

Ruffins, S.V. and Ch.A. Ettensohn (1996) A fate map of the vegetal plate of the sea urchin (*Lytechinus variegatus*) mesenchyme blastula, *Development* **122**, 253-263.

Ryabova, L.V. (1995). Two-component cytoskeletal system as a basis of cortical contractility in *Xenopus laevis* eggs. *Ontogenez (Russ.J.Devel Biol.)* **26**, 236-247.

Ryan, K., Garret, N., Mitchel, A., Gurdon, J.B. (1996) Eomesodermin, a key early gene in *Xenopus* mesoderm differentiation, *Cell* **87**, 989-1000.

Saxen, L., Toivonen, S. (1962). *Primary Embryonic Induction*. Academic Press, N.Y., L.

Schroeder, E. (1990). The contractile ring and furrowing in dividing cells. *Ann. N.Y. Acad. Sci.* **582**, 78-87.

Selman, G (1958) The forces producing neural closure in Amphibia, *J. Embryol. Exp. Morphol.* **6**, 448-465.

Shih, J. and R. Keller (1992) Patterns of cell motility in the organizer and dorsal mesoderm of *Xenopus laevis*, *Development* **116**, 915-930.

Shimizu, H. (1984) An actomyosin motor, in G.H. Pollard (ed) *Contractile mechanisms in muscle*, Plenum Press, New York, London, pp. 429-436.

Shubin, N., Tabin, C., and Carrol, S. (1997) Fossils, genes and the evolution of animal limb, *Nature* **388**, 639-648.

Shubnikov, A.V., Koptzik, V.A. (1972) *Symmetry in Science and Art*. Nauka, Moskva (in Russian).

Shyy, J. Y-J., Chien, Shu (1997) Role of integrins in cellular responses to mechanical stress and adhesion, *Curr. Op. Cell Biol.* **9**, 707-713.

Singhvi, R., Kumar, A., Lopez, G.P., Stephanopoulos, G.N., Wang, D.I.C., Whitesides, G.M. and Ingber, D.E. (1994) Engineering cell shape and function, *Science* **264**, 696-698.

Skibbens, R.V. and E.D.Salmon (1994) Kinetochore directional instability in vertebrate mitotic cells, in N. Akkas (ed.) *Biomechanics of Active Movement and Division of Cells* (NATO ASI Series H: Cell Biology), Springer Verlag, Berlin, Heidelberg, pp. 545-550.

Sinnot, E, W. (1960) *Plant Morphogenesis*, McGraw-Hill Book Company, N.Y., L.

Slack, J.M.W. (1994). Inducing factors in Xenopus early embryos. *Current Biology* **4**, 116-126.

Slawinski, J. (1988). Luminescence research and its relation to ultraweak cell radiation. *Experientia* **44**, 559-571.

Smith, J.C. (1993). Mesoderm-inducing factors in early vertebrate development. *EMBO J.* **12**, 4463-4470.

Snape, A., C.C. Wylie, J.C. Smith, J. Heasman (1987) Changes in states of commitment of single animal pole blastomeres of *Xenopus laevis, Devel Biol.* **119**, 503-510.

Sokol, S.I. (1994) The pregastrula establishment of gene expression pattern in *Xenopus* embryos: requirements for local cell interactions and for protein synthesis, *Devel Biol.* **166**, 782-788.

Solnica-Krezel, L., D.L. Stemple and W. Driever (1995), Transparent things: cell fates and cell movements during early embryogenesis of sebrafish. *BioEssays* **17**, 931-939.

Spemann, H. (1936). *Experimentelle Beitraege zu einer Theorie der Entwicklung.* Springer, Berlin.

Spirov, A.V. (1989). Some theoretical aspects of a biological morphogenesis: interrelation between the pattern and the shape. *Zhurn. Obsch. Biol.* **50**, 606-620 (in Russian).

Steding, G.(1967).Ursachen der embryonalen Epithelverdickungen. *Acta Anatomica* **68**, 37-67.

Steinberg, M.S. (1978). Cell-cell recognition in multicellular assembly: levels of specificity. In: *Cell-Cell Recognition* (A.S.G.Curtis ed.), pp. 25-49. Cambridge Univ. Press, Cambridge.

Stern, C.D. (1984) A simple model for early morphogenesis, *J. theor. Biol.,* **107**, 229-242.

Stern, C.D. and MacKenzie, D.O. (1983). Sodium transport and the control of epiblast polarity in the early chick embryo. *J. Embryol. Exp. Morphol.* **77**, 73-98.

Stopak, D. and A.K.Harris (1982). Connective tissue morphogenesis by fibroblast traction. *Devel Biol.* **90**, 383-398.

Strohman, R.C. (1997). The coming Kuhnian revolution in biology. *Nature Biotechnology*, **15**, 194-200.

Svetina S. and B. Zeks (1990) The mechanical behavior of cell membranes as a possible physical origin of cell polarity *J.theor.Biol.* **146**, 115-122.

Svetina, S. and B. Zeks (1991) Mechanical behavior of closed lamellar membranes as a possible common mechanism for the establishment of developmental shapes. *Int.J.Dev.Biol.* **35**, 359-365.

Symes K., Smith, J.C. (1987) Gastrulation movements provide an early marker of mesoderm induction in Xenopus laevis. *Development* **101**, 339-349.

Taber, L.A., I.-en Lin and E.B.Clark (1995) Mechanics of cardiac looping , *Devel Dynamics* **203**, 42-50.

Thompson, D'Arcy (1942) *On Growth and Form*, Cambridge University Press, Cambridge.

Toshiaki, Iba, Sumpio Bauer E. (1992) Tissue plasminogen activator expression in endothelial cells exposed to cyclic strain in vitro. *Cell Transplant.* **1**, 43-50.

Traub, P. and Shoeman, R (1994) Intermediate filament proteins: cytoskeletal elements with gene-regulatory function? *Int. Rev. Cytol.* **154**, 1-104.

Trinkaus, J.P. (1978) Mediation of cell surface behavior by intercellular contacts. *Zoon* **6**, 51-63.

Turing, A.M. (1952) The chemical basis of morphogenesis. *Phil. Trans. Roy. Soc. L, ser B*, **237**, 37-72.

Vaishnav, R.N., Vossoughi, J. (1987), Residual stress and strain in aortic segments.
   J. Biomechanics **20**, 235-239.

Vasieva, O.O. (1991) *Study of the processes of autowave aggregation in the population of Dictyostelium discoideum cells.* Candidate theses. Inst. of experimental and theoretical biophysics, Russian Acad. Sci. Pushino (Russ).

Vasiliev, V.A., Romanovsky, Ju.M., Chernavsky, D.S. (1982). Elements of the dissipative structures theory: a relation to the structuration problem, in A.I. Zotin, E.V.Presnov (eds) *Mathematical Biology of Development*, Nauka, Moskva, pp. 82-101 (Russ).

Verdonk, N.H. (1968) The determination of bilateral symmetry in the head region of *Limnaea stagnalis, Acta Embryol. Morphol. Exp.* **10**, 211-227.

Vodicka, M.A. and J.C.Gerhart (1995) Blastomere derivation and domains of gene expression in the Spemann organizer of *Xenopus laevis, Development* **121**, 3505-3518.

Waddington, C.H. (1940) *Organisers and Genes*, Cambridge University Press, Campbridge.

Waddington, C.H., Yao, T. (1950) Studies on regional specificity within the organizational center of the Urodeles, *J. exp. Biol.* **27**, 126-144.

Wanek, N., Marcum, B.A., Lee, H.T., Campbell, R.D. (1980) Effect of hydrostatic pressure on morphogenesis in nerve-free hydra, *J. exp. Zool.* **211**, 275-280.

Watterson, J.G. (1991) The role of water in cell functioning, *Biophysika* **36**, 5-30 (in Russian).

Weyl, H. (1952) *Symmetry.* Princeton University Press, Princeton.

White, J.G. (1990) Laterally mobile, cortical tension elements can self-assemble into a contractile ring. *Ann. N.Y.Acad. Sci.* **582,** 50-59.

White, J.G. and G.G.Borisy (1983) On the mechanisms of cytokinesis in animal cells, *J. Theor. Biol.* **101,** 289-316.

Wilson, P.A. and D.A.Melton (1994) Mesodermal patterning by an inducer gradient depends on secondary cell-cell communications, *Current Biology* **4,** 676-686.

Winfree, A.T. (1991) Crystals from dreams, *Nature* **352,** 568-569.

Wirtz, H.R.W. and L.G. Dobbs (1990) Calcium mobilization and exocytosis after one mechanical stretch of lung epithelial cells, *Science* **298,** 1266-1269.

Wohlfarth-Bottermann, K.-E. (1987) Dynamic organization and force production in cytoplasmic strands, in J. Bereiter-Hahn, O.R.Anderson and W.-E. Reif eds, *Cytomechanics*, Springer Verlag, Berlin, Heidelberg, pp. 154-168.

Wolpert, L. (1996) One hundred years of positional information, *Trends in Genetics* **12,** 359-364.

Wylie, C.C., A. Snape, J. Heasman and J.C. Smith (1987) Vegetal pole cells and commitment to form endoderm in *Xenopus laevis, Devel. Biol.* **119,** 496-502.

Wyttenbach, C.R., Crowell, S., Suddith, R. (1973) Variations in the mode of stolon growth among different genera of colonial hydroids, and their evolutionary implications, *J. Morphol.* **139,** 363-375.

Yamada, T. (1981) The concept of embryonic induction in the passage of time, *Netherlands J. of Zool.* **31,** 78-98.

Yamada, T. (1994) Caudalization by the amphibian organizer: *brachyuri,* convergent extension and retinoic acid. *Development* **120,** 3051-3062.

Yermakov, A.S. and Beloussov, L.V. (1998) Variability and asymmetry of the axial rudiments in *Xenopus laevis* embryos after the disturbance of cellmovements and tensile fields in the marginal zone of early gastrulae, *Ontogenez (Russ.J.Devel Biol.)* **29,** 38-46.

Yost, H.J. (1991) Development of the left-right axis in amphibians, in *Biological Asymmetry and Handedness.* Wiley, Chichester (Ciba Foundation Symposium 162) pp. 165-181.

Zacharov, V.M. (1987) *Asymmetry in Animals.* Nauka, Moskva, (Russ).

Zaraisky, A.G. (1991) Self-organization during determination of the size of axial structures in Xenopus laevis embryos, *Ontogenez (Sov. J. Devel. Biol.)* **22,** 365-374.

Zotin, A.I., Zotina, R.S. (1993) *Phenomenological Theory of Development, Growth and Ageing.* Moskva, Nauka (in Russian, with English summary).

INDEX

DATE DUE